FORAGING THE LAND

A Comprehensive Beginners Guide to Identifying, Harvesting, and Preparing Edible Wild Plants

EDDIE HOLDEN

EDDIEHOLDEN.COM

Contents

A FREE GIFT TO OUR READERS

Thanks for buying this book! Here is a list of 5 foolish mistakes foragers make and how to avoid them that you can read right now! Scan the QR code below for your free copy!

http://eddieholden.com

Introduction

Foraging is one of the oldest skills in human history. But does it have a place in modern life? The short answer, which shouldn't surprise someone reading a foraging guide, is "Yes."

Some might assume that people only forage as a last resort, perhaps because they're short on cash and can't get all the groceries they need. However, most people who forage do so because they pride themselves on being self-sufficient, responsible, and resourceful. As well as this, the natural world has been proven to have a beneficial effect on people's well-being, reducing stress and encouraging a healthier lifestyle. There's no such thing as being too healthy or stress-free, so why not include foraging in your daily routine?

Common Foraging Concerns

It's perfectly common to be somewhat worried about certain aspects of foraging, but it does help to identify these issues. Perhaps these concerns hold you back from foraging, or they're what drive you to want to learn more about it. Once you know what your issues are, you can understand them and, hopefully, do something about them. Here are some of the most common concerns that you might have:

1. Misidentification of plants. This is the big one and for a good reason. The natural world is full of gifts and wonders, but there are a few more problematic plants out there that you shouldn't consume. If you eat something that you shouldn't eat, it can be unpleasant or even dangerous.

2. Lack of confidence. Whenever anyone picks up a new skill, they have that moment of worry that they don't know what they're doing. We've all been there, whether it's foraging, cooking, or even parenting. You might be asking yourself questions like:

"Do I have the right gear or equipment?" Am I harvesting plants correctly?" Where can I find an abundance of edible plants?"

3. Unreliable or insufficient foraging resources. On a similar note, you may feel that some of the foraging resources and information out there simply isn't good enough. Maybe you can't trust the foraging resources that you have available. Or perhaps they don't contain enough detailed information for you to feel confident enough to go foraging.
4. Possibility of getting stranded and not knowing how to survive. While foraging isn't just a tool for when you're stuck in the wilderness, it can be a vital part of survival. If you get stuck in the wild, you may be concerned that you won't know what plants to look for or how to distinguish between edible and inedible lookalikes.

Those concerns all make a fantastic catalyst for buying this book. Let's face it; nobody buys a book based entirely on its title or cover. When you find a book like this one, it's because you're looking for a helping hand. I've been around the block a few times, but I understand the drive to learn more about something. For most readers, you want to be able to confidently and safely forage plants, which you'll be able to do once you learn the skills and knowledge necessary through this guide.

Who Can Learn

One of the great things about foraging is that you can't be too young or old to learn. This makes it a brilliant activity for families and single adults alike. I've been foraging for almost as long as I can remember, so I have a few decades of experience behind me. But, just like you, I had to start somewhere. I'm certainly not a novice anymore, but I'm still learning. I keep picking up new tricks and ways to use what I find.

While foraging isn't as popular as it used to be, you can still find examples of young people venturing into the wilderness to find some delights. Some teenagers document their foraging experiences through TikTok[1], which means that foraging and sustainable living are being slowly but surely brought into the mainstream. If you've got kids or teenagers and

[1] (Retta, 2021)

want to encourage them to spend more time in the great outdoors, then finding social media accounts like these can be a great way to educate and encourage them.

I'm going to talk about why foraging is so important in more detail later on, but I can't stress enough how valuable this skill is. Once you've gone through this guide and taken on some of the advice within, you will be confident enough to see the results for yourself. The great thing is that these skills fit in with every lifestyle. For now, let me paint a picture or three.

You're out and about; perhaps you're wild camping or on a hike. You may even be in a survival situation. You spot something that you recognize as an edible plant. You're confident in your ability to harvest it safely and know precisely what it is. You can either eat it right away or prepare it as part of a campfire meal. Nature is working for you.

You take your family out berry picking. You know what you're looking for and where to find it. You can teach your children how to spot the berries and how tasty they are. You take plenty back for jam or pie or simply as healthy treats for the whole family.

You're planning on cooking a meal, and you know that you can find fresh, wild produce nearby. Perhaps you've already harvested it or even preserved it, which means that you can add a new dimension to your food. You know that fresh, unprocessed food is the healthiest option, and foraging has given you an affordable way to find it. You'll feel better for doing it, and you've even saved a few bucks in the process.

It sounds nice, doesn't it? Is it a bit unrealistic? Maybe a bit too good to be true? The short answer is no. You'll have to read the rest of the book to find the long answer.

How This Book Can Help

If you've looked at the front page, you know my name. But you might be wondering, "why listen to this guy?" Which is fair enough.

As I've mentioned, I've been foraging since I was a child; I remember tagging along with my parents as they'd find all kinds of edible wonders out in nature. We'd either eat them on the spot, or my parents would cook them into extraordinary meals. I remember the food about as fondly as the foraging.

Since then, that love and passion for foraging have never left me. I've bounced around a few states across North America and gained even more knowledge as I've foraged in the many different climates and locations. Over the years, I've gathered an extensive store of knowledge and skills to find wild edible plants and turn them into something extraordinary.

My parents gave me an education and an opportunity to learn that I'll always be grateful for, and all I want to do now is return the favor. I'm teaching my children how to follow in their footsteps, and I don't want this skill and knowledge to die. So, that's why I've written this book. I'm confident that everything that I've learned will be valuable to a beginner forager such as yourself, and it's great to be able to help you embark on a foraging adventure of your own.

Experience is a great teacher if sometimes a cruel one. For example, it took me a few tries to get the hang of harvesting stinging nettles without getting stung when I first tried, but all the nettle tea I've enjoyed has made it more than worth that initial discomfort. But even better, I'm able to pass on that experience without you having to go through quite the same trials and tribulations.

Foraging is affordable and attainable as long as you have the proper guidance. Do you remember the four common concerns I mentioned before? Without good advice and the correct information, those issues will make it far more complicated than it needs to be to get into foraging. There are so many easily avoidable mistakes and pitfalls that overcomplicate foraging, which this book will help you to avoid. You can maximize your harvest and minimize foraging faux pas.

With that being said, it's time for you to find out for yourself how practical this advice is. Read on and put it into practice, and keep those mental pictures of your potential future in mind.

Speaking of pictures, the QR code on the next page links to a document containing helpful diagrams and photographs of plants and mushrooms that you can forage. Keep it handy to compare it to what you can find in the wild.

SCAN ME

1

The Basics of Foraging and Why Foraging Matters

Henry David Thoreau, a man who I'm not afraid to say is far wiser and better with language than I am, had his thoughts on nature. He said that "we need the tonic of wilderness… At the same time that we are earnest to explore and learn all things, we require that all things be mysterious and unexplorable, that land and sea be indefinitely wild, unsurveyed and unfathomed by us because unfathomable. We can never have enough of nature."

What do the words of a philosopher who famously hid away in the woods have to do with foraging? Well, he truly came to understand the benefits of living off the land, a large part of which includes foraging. But before plunging into the details of why people should take up foraging and how you can do so, you should get a basic definition and understanding of foraging under your belt.

What is Foraging?

A basic description of foraging is to search for, identify, and collect wild, uncultivated food resources.[2] You can find these wild edible foods in rural and urban environments alike. So,

---

[2] (Docio, n.d.)

you city-dwellers needn't feel entirely left out. You just need to know where to look. As you'd expect, this food is all free and entirely unprocessed.

There are several degrees of foraging. Depending on what suits you more, you can integrate it into your lifestyle however you choose. Some people only forage for berries on rare occasions, which is an enjoyable experience for the whole family. However, foraging more extensively can enrich your lifestyle far more. Even if you don't want to cut out processed foods entirely, you can change the way you eat so that you don't have to rely on them.

Why Foraging Matters

Humans have foraged for food since the dawn of time. True, they foraged out of necessity, and modern humans have other options. But that's no reason to relegate foraging to a footnote in a history book. Arguably, this connection to our ancestors is one reason why some people fall in love with foraging. This desire to live off the land and use what surrounds us is almost instinctual. Human history is a part of every individual, and the skills that we have developed over the years have kept us alive and have allowed us to form societies. They deserve more than to fall by the wayside.

That is, admittedly, a somewhat romantic view of things. Sure, foraging is important because of its place in history, but it can also be a practical part of modern life. The benefits of foraging can be broken down into six parts so that you can see for yourself how it can help you, your family, and the world around you.

1. Financial Benefits

Money, for better or for worse, makes the world go round. Many of us are preoccupied with saving money to protect our futures and, hopefully, live a comfortable life. When it comes to finances, every little helps.

This means that chances are, the opportunity to save money[3] is one of the more compelling reasons to get into foraging. It's no secret that humans need food to survive. Unfortunately, food prices are constantly rising, and the cost of plants is even more so. This rise is noticeable worldwide, but especially in the United States.

---

[3] (Deane, 2011)

This "food inflation" affects us all. Foraging can be a helpful way to supplement your weekly grocery shopping, meaning that you don't need to buy everything at the store. Instead, you can fill in the gaps with foraged food, potentially saving plenty of money. Even if you can't find edible plants every week, cutting your costs never hurts.

2. Environmental Benefits

While money might make the world go round, and saving money can help secure your future, there's something else to consider. To put it bluntly, all of the money in the world won't help if there isn't anything left of it. Foraging on its own isn't going to halt every environmental issue hitting the planet, but it's something that each of us can do to help.

When you forage food, you immediately cut down on the amount of plastic used in food packaging. Wild plants are also seasonal and grow nearby, so your fruit and vegetables aren't flown from the other side of the world. The fewer steps involved in getting the food from the ground and onto your plate, the better it is for the environment.

Not only that, foraging food takes effort. It's enjoyable, but it's still something that you have to work at. That means you're far less likely to let your hard work go to waste by allowing the foraged food to spoil. Instead, you'll have yet another incentive to use or preserve your food, reducing food waste. Foraging is one significant step towards following a more sustainable and environmentally conscious lifestyle.

3. Physical Benefits

It's no secret that fresh fruits and vegetables typically come packed with nutrients. While we might be aware of the health benefits of cultivated plants such as kale (everyone's favorite go-to health food), you can find the same nutrients in wild plants.[4]

For example, the common dock leaf is considered an invasive weed in the United States. But it boasts an impressive amount of vitamin A and C. Only 100 grams (3.5 ounces) of the stuff can provide 80% of your daily allowance of these vitamins. Also, once you've foraged fresh food, it's up to you to prepare and eat it. Processed foods and takeout are convenient and tasty, but home-cooked food is typically far more healthy as it doesn't contain anywhere

---

[4] (Mabey, 2015)

near as much sugar, fat, or salt. Foraging vegetables encourages you to cook them at home, so you end up eating this nutritious fresh food instead of the unhealthier options.

Centuries ago, people used foraged plants as medication. Even though they couldn't identify the exact health benefits of a plant, they recognized that these plants were good for them. You can use the same plants to help your health, as long as you do your research and ensure that they won't interact with any medications or other health conditions. For example, you can whip up a quick tea for a great vitamin boost.

As well as the health benefits of eating this wild food, foraging is an inherently active hobby. You have to get out there and find wild plants. While most foraging isn't exactly high-intensity exercise, it does encourage you and your family to get out there and walk around. Even half an hour of light exercise is better than nothing, but you can also integrate foraging into longer hikes. If you're able to identify plants along your route, then there's nothing stopping you from grabbing a quick natural snack along the way. The two activities go hand-in-hand.

4. Mental Benefits

Going outside and foraging isn't just good for your physical health, but it can also do wonders for your mental health. Humans crave the outside world, but modern life has a habit of getting in the way. But when you decide to set aside some time to go foraging, you have to go outside and find a natural spot for the best foraging opportunities.

The outside world has proven benefits to our physical and mental well-being. Simply being out in the sun can raise our vitamin D production, which is difficult to get from other sources. Vitamin D[5] strengthens the immune system, improves bone health, and can boost our mood. It's not known as "the happiness vitamin" without good reason. Studies have suggested that vitamin D deficiency can contribute to poor mental health and physical conditions. Spending time outside anywhere can benefit our mental health, but being among nature and plant life is even more beneficial. [6]

---

[5] (Brennan, 2021)

[6] (*How Nature Benefits Mental Health*, n.d.)

5. Social Benefits

True, it is possible to forage alone, but foraging is an inherently social activity. I've mentioned before that it's a great way to spend time with your family and specifically your children, but you can forage with anyone you like. As foraging is an inexpensive hobby that gets you outside, it's a perfect way to spend time with someone without breaking the bank.

Foraging also provides an opportunity to get to know new people. Chances are, there's a group of like-minded people nearby who are either interested in getting started with foraging or who have been doing it for a while. So, look for these groups online, and you never know who you might meet. If you find someone more experienced than you, you can benefit from their knowledge in this field. But even if you go foraging with a novice, two people can spot more wild plants than one. Even better, you can carry more of your treasures back with you.

6. Flavor Benefits

All of the benefits listed above are great reasons to get out there and get foraging. But you can't deny the fantastic flavor opportunities you're opening yourself to. One of the major issues with buying groceries is that much of what you purchase is out of season.[7] This has several ramifications. First of all, it's often more expensive and has to be shipped from miles away, which has a knock-on effect on the environment. But another issue is that it doesn't taste as good as fresh, seasonal food.

If you haven't eaten according to the seasons before, you've been missing out. Fresh fruit and vegetables that are perfectly in season are bursting with fantastic flavors that you can't replicate with anything else. You can create meals that cater to the fussiest palate when cooking with these ingredients. One of my fondest memories of foraging is the incredible food my parents created using the treasures we had discovered earlier.

When you go foraging, you can find ingredients that aren't as easy to locate in the stores. Even if you can buy them, they're usually so much better when you find them yourself. This revitalizes your cooking experience. Cooking can, sadly, turn into a bit of a chore when you find yourself churning out the same meals every week. But if you throw foraging into the mix, it's inherently more exciting, and every meal is slightly different. Foraging invites

[7] (Baker, 2021)

creativity in the kitchen, which can inject some fun into what would otherwise be a daily chore. I'm not saying that you have to create every meal entirely from foraged ingredients, although that would be a fun challenge; switch a few things out now and then, and you never know what wonders you'll invent.

When cooking with foraged ingredients, it's important to taste them before throwing them into the pot. You want to know what flavors you're working with so that you can get the most out of them. It might seem daunting to start with, but it soon becomes second nature, and you will be able to taste the difference on the plate.

What Does it Really Mean to Be a Forager?

So, we know the definition of foraging. We know some of the benefits of foraging and why people choose to do it even though you can buy food from the grocery store. But what does it mean to be a forager?

The dictionary definition of a forager is "a person or animal that searches widely for food or provisions." But while that definition is accurate, there's more to it than that.[8]

A forager is independent and self-sufficient. They can fend for themselves using the land, so they don't have to rely on the grocery store. A forager doesn't have to be a survivalist, but they can use the wilderness to help them to survive. A forager is a steward of nature rather than being disconnected from the natural world around them. A forager says no to the stressful way of life that's all too common in this modern world, where you're too busy to breathe, but it never feels as though you're going anywhere.

Anyone can be a forager. You just have to step outside.

Before You Start: Foraging Basics

But before you venture into the great outdoors, it helps to know what you're doing. As I've said, this book will give you the knowledge you need to forage safely and get the most out of what you harvest. But first, here are the basics.

---

[8] (*What Is Foraging?*, 2021)

Tools and Gear for Foraging

At a stretch, you can forage berries with your hands and snack on them on the go. But if you wanted to forage other plants, or if you wanted to take some of your prizes home with you, then you might need a little bit of equipment.[9] Don't worry; foraging is still an inherently cheap hobby, and you don't need anything too extravagant or expensive. There are even more options for equipment than described here, but these are the essentials.

- **Cutting Tools.** You'll want some basic items to help you forage plants more efficiently. These might include good-quality scissors and a pocket knife. A trowel or fork can also help you forage roots.
- **Gloves.** If you're foraging for a while, you'll soon notice that your hands get muddy and scratched up by thorns, branches, or stinging nettles. Some sturdy gloves will protect your hands.
- **Baskets or Bags.** It would be best if you had something to bring your harvest back with you. A classic wicker basket is a nice touch, but you can also use a more practical mesh bag to carry your plants easily.
- **Protective Clothing.** As well as your gloves, it's beneficial to wear other clothes to protect yourself. Good boots will make it easier to reach more wild areas, while a long-sleeved shirt will protect you from poison ivy or other irritants. A hat will keep the sun off your face, and long pants can protect you from ticks.
- **Field Guide and Magnifying Glass.** When foraging, it's essential to make sure that you know exactly what you're picking. A pocket-sized illustrated field guide is a must, and modern technology allows you to keep plenty of information on your phone as well. Ideally, you should have a few different data sources to be as safe as possible. If you're keeping the guides on your phone, then download them before you go out foraging so that you don't get stuck waiting for a poor signal.
- **Handheld Magnifying Glass or Loupe.** A magnifying glass can help you to identify a plant up close. You could also invest in an illuminated metal loupe, a portable magnifying glass with an attached light. Use this tool, along with a foraging guide, to figure out precisely what you're looking at.

---

[9] (Blankespoor & Gemma, 2021) (*4 Essential Foraging Tools*, 2019)

FORAGING THE LAND

Where to Forage

The wonderful thing about foraging is that you can do it in pretty much any green space. Even people in urban areas can often find somewhere nearby to forage, although it's generally better to go to a spot where there are fewer people and especially vehicles around. You don't have to travel too far when you first venture out. If you own or rent a property with land (such as your yard), start there. Look for green spaces that could hold edible plants. You might be surprised at what you can find only a few yards from your back door.

You have even more options if you live in a rural area or near a national park. However, it's essential to be aware of the local laws[10] in your area. If there's a promising piece of land nearby, then make sure that you ask the property owner for permission before wandering around and picking plants. You should also ask if they use any toxic pesticides or other chemicals so that you don't risk eating contaminated food.

One word of caution, however. Chemicals, pesticides, pollutants, or other toxins can contaminate some areas. Avoid land near landfills, factories, train tracks, busy roads, or other places that are likely to be contaminated. Even if the plants look healthy, they could make you unwell.

Depending on where you live, you might find a wide variety of plant life. In the United States alone, I've explored several different environments. North America has marshland, desert, temperate grassland, tundra, and many other biomes. All of these areas are best suited for different kinds of plant life. It's one of the reasons that I find foraging so exciting; I never know what I'll stumble upon next. The variety of climates and flora expand even more around the world, so it's essential to do research specific to your area. Even if you're in the same country, there's no guarantee that you'll find the same plants everywhere.

As a general rule, you can find plants to forage in all kinds of different environments. Grasslands and plains feature all sorts of wonderful weeds, mushrooms, and shrubs to choose from. Woodland is full of trees where you can find fruit or nuts, bushes, and plenty of other plants. Hedgerows are often a rich source of berries. You've found a good spot when you find a group of edible plants in abundance. Just stay safe and ask permission if you need to.

---

[10] (*Foraging Legality — Four Season Foraging*, 2019)

When to Forage

You can find wild edible plants throughout most of the year.[11] As you'd expect, specific varieties of plants are more prevalent at different times. You'll be pleased to hear that there's always something growing out there, even during the colder months.

Along with the field guide mentioned above, it's a good idea to use a foraging calendar to find out what grows when in your area. You can find monthly foraging calendars[12] for your country and region to help you know what to look for when you're out. Depending on your climate, different plants may be available in different months. Even within the same country, you might find that plants that grow in March in one area won't be ready to harvest until April further north.

How to Forage

This guide will provide you with more detailed information later on, but there are some basic tips and principles to foraging that you should always bear in mind.[13]

- Safety comes first. Again, there will be a whole chapter devoted to staying safe while foraging, but it's a point that's worth repeating. Never eat anything that you aren't sure of. There are a lot of delicious wild plants out there that are perfectly edible, but there are also some that will make you very ill.
- Start with easy species that are easy to identify. Some of the tastiest wild edibles are also very common and easy to distinguish from other plants. These are great starting points for any forager.
- Don't overharvest. Yes, it's exciting when you find a place that's rich with plenty of foraging opportunities, but don't strip the place bare. Ideally, the area shouldn't look too different after leaving it. If you take too much, then the local wildlife will suffer, and the wild food might not return in such abundance next year. Remember, one of the reasons to forage is to help the environment, so just take what you need; you can always come back later.

---

[11] (*Monthly Foraging Guide: What's in Season, Where to Find It, and How to Forage Responsibly*, n.d.)

[12] (Huffstetler, 2019)

[13] (TyrantFarms, 2019)

- Harvest carefully and properly. If you're harvesting berries, don't destroy half of the branch in the process. The same applies to leaves and other parts of the plant. You don't want to kill the main plant, so even when you're harvesting roots, do it with care. This way, you know that it'll return, and your harvest will be of a higher quality.
- Keep a journal. Some plants have a habit of returning to the same area, especially if you harvest responsibly. So keep a journal of what you find where, so you will eventually have a guide of your own that's specific to your local area.

To Conclude

As with any undertaking or project, it's always important to be clear about the essentials first when it comes to foraging. This will increase your chances of having a valuable foraging experience, with plenty of success in the future.

Now that you have the basics in hand and, hopefully, you're more excited about potentially becoming a forager, it's time to move on to the next piece of the puzzle. The next chapter will help you grasp a basic but no less crucial understanding of botany. Botany is a vital part of your foraging adventures and will help you safely and effectively start your journey.

2

Botany Essentials for Foragers

It's no secret that knowledge is power. That adage applies to nearly everything, including becoming a safe and successful forager. After all, if you're going to put anything in your mouth, it's essential to know precisely what it is. In order to forage productively and responsibly, you need an understanding of botany[14]. That might sound a little intimidating. Do you really need a degree in botany to learn how to forage?

In short, no. You might also be skeptical about learning botany because it sounds a bit like going back to school. For some people, that's a dream come true. For others, it's more likely a nightmare. John Burroughs points out that "Most young people find botany a dull study. So it is, as taught from the text-books in the schools; but study it yourself in the fields and woods, and you will find it a source of perennial delight."

The whole point of this chapter is to help you get out there into the fields and the woods so that you can practically apply the basic principles of botany. But before you can use botany in the field, you need to at least understand what it is and how it can help you. Remember, foraging is, by definition, one of the most accessible hobbies to get into. Millennia ago, gatherers who foraged to survive didn't have to spend years learning the ins and outs of botany. However, knowing what you're doing gives you an edge and can help you avoid the unfortunate mistakes that past foragers made. You can always study more deeply if you take an interest in it, but you don't need to become an expert to be able to forage. However, you

---

[14] (VanDerZanden, n.d.)

will be able to accurately identify the plants that you want and know how to tell how mature they are and which parts are best to eat.

General Botany Concepts Explained

As with any subject, the easiest way to learn about botany is to break it down into the main concepts. Each of these concepts will help you on your foraging adventures, and I'll give you an overview of what each of them means. Once you know what the concepts mean, you can use them to help you to identify the plants that you encounter.

Please keep the QR code mentioned in the introduction handy as you go through this chapter. There, you will find diagrams to explain the following concepts further.

The Life Cycle of Plants

The life cycle of a plant is, as you'd expect, a look at the different stages of a plant's life. Every plant will have the same stages, but they might look different for different plants or take a different amount of time. The basic stages are as follows:

- Seed germination, where the seed first starts to sprout.
- Seedling, where you can view the first visible greenery. This is usually a short stem and a couple of tiny leaves.
- Vegetative growth, where the stem gets taller and may branch off, and leaves grow.
- Flowering, many plants produce flowers that allow insects and other pollinators to spread their pollen and fertilize other plants (and vice versa).
- Pollination/seed production, once pollinators fertilize the flowers, they will produce new seeds. This may take the form of fruit or berries, which animals who eat the fruit will spread. The seeds are then dispersed (by wind or animals), and a new plant can grow.

Plant Growth and Development

Three significant functions drive plant growth and development. These functions are necessary for survival, much like humans need to eat, breathe, and drink to survive. In plants, these functions are photosynthesis, respiration, and transpiration.

- Photosynthesis is the process by which plants manufacture their food. The plant uses energy from sunlight, along with carbon dioxide and water, in a chemical reaction to produce glucose and oxygen.
- Respiration, as in animals, is the process where the plant converts glucose into energy. In some respects, it's the opposite process of photosynthesis, where glucose and oxygen combine to produce carbon dioxide, water, and energy. However, the energy produced is in a form that the plant can directly use.
- Transpiration allows a plant to "drink". It occurs when a plant releases water vapor through its leaves, which causes more water to be pulled into and through the plant via the roots. This allows minerals and other chemicals to be transported through the plant via water and allows plant cells to stay hydrated. It also cools the plant as the water evaporates on the leaves.

Each of these processes is carefully balanced so that the plant can thrive. When circumstances are ideal and these processes are working correctly, the plant can continue growing and developing. The environment can impact a plant's ability to photosynthesize, respire, or transpire. For example, if a plant can't transpire because of a lack of water, its ability to photosynthesize is compromised. This, in turn, means that the plant doesn't have the glucose needed to respire and grow.

The Structure of Plants

It should be increasingly evident that the structure of a plant has a dramatic effect on its ability to function. A plant is made up of both internal and external structures. Knowing what these structures are and how they work will allow you to harvest a plant without causing any irreparable damage.

Internal Plant Parts

Plants are made up of cells, which take care of most plant reactions. So, photosynthesis, cell division, and respiration all occur at the cellular level. Larger plant tissues are made up of organized groups of similar cells.

Plant cells are unique in that almost all of them contain all of the genetic material required to develop into a whole plant. This is why it's possible to take plant cuttings and grow them into a new plant. It doesn't mean that if you pull off a leaf and drop it on the ground, a whole new plant will always grow from that spot. But it does allow for some exciting grafting opportunities. With some care, you can develop a new plant from a cutting or even transplant it onto a different plant. But I digress.

Plants also have specialized groups of cells that are particularly interesting. They're called meristems, and they are where the majority of plant growth occurs. They're important because you can manipulate how meristems act, allowing you to change a plant's growth pattern, make it branch out more, or encourage flower and vegetable growth. But this is something that gardeners will take more seriously, as foragers don't need to interfere with wild plants in the same way.

External Plant Parts

However, foragers should pay attention to external plant parts or the organs of the plants. An organ, as in humans, is made up of an organized group of tissues that work together to perform a specific function. Different plant parts are more or less suitable for eating, depending on which plant you harvest. However, you will have likely come across or eaten all of these parts.

Here are the external plant parts that you'll come across. They are often split into two categories, which are sexual reproductive parts and vegetative parts. Bear in mind that this section is specifically about flowering plants, as some don't produce flowers or fruit. We'll mention those soon enough, don't worry. Sexual reproductive parts:

- Flower buds
- Flowers
- Fruits
- Seeds

These parts of the plant are responsible for reproduction. The flower buds grow into flowers, which produce pollen. Pollinators, such as insects and small birds, spread this pollen. When the pollinators go to other flowers, they fertilize them with the pollen from the last flower. This, in turn, causes the flower to develop fruit. There are several types of fruits, but they're all designed to spread the seeds.

Finally, the seeds start the whole process over again. When the fruit is eaten or otherwise moved away from the original plant, the seed can grow into a new plant. Every seed contains all the genetic information required to develop; it just needs the right conditions to germinate. A seed will only germinate under these conditions so that the plant has more chance of surviving into adulthood. Until then, it can remain dormant for years, although older seeds are less likely to be viable as time goes on.

Vegetative parts:

- Roots. While the roots are less visible than other parts of the plant, they are critical to its health. A root system will often be pretty large, which is important as the roots collect water and nutrients from the soil. The roots also keep the plant secure in the ground and bind the soil together.
- Stems. The stem makes up much of the structure of a plant above ground. It supports buds and leaves and carries water, minerals, and food to wherever it needs to go.
- Shoot buds. A bud is a shoot that can either grow leaves or flowers. Leaf buds tend to be more pointed and thinner than flower buds. The bud form is hardier than flowers or leaves and will only continue to develop when the environment is suitable.
- Leaves. The primary function of a plant's leaves is to absorb sunlight to allow photosynthesis to occur. A plant will also transpire through its leaves.

Every vegetative organ has a vital function, which means destroying one organ can destroy the plant. But if you're careful, you can harvest it safely.

How Plants Are Classified

Botany is essential to the budding forager because it allows you to identify plants correctly. You can classify plants in various ways, and each classification narrows down the type of plant you're looking for. Knowing what different plant classifications are can help you to identify them.

Formal Classifications (Taxonomy)

You can identify every living organism using a system called Biological Classification, or Taxonomy[15]. This is where all those difficult-to-pronounce Latin scientific names come from.

---

[15] (*Basic Botany*, n.d.)

Taxonomy uses physical characteristics and genetics to organize organisms. One popular arrangement of this method is as follows:

1. Kingdom
2. Phylum
3. Class
4. Order
5. Family
6. Genus
7. Species

So, what does this mean? Well, each classification narrows down an organism into a different group. For example, dogs and gray wolves are different species, but they share a genus. This means they're more closely related than humans and gorillas, who only share a family. All animals are grouped into the same kingdom (Animalia), but plants and mushrooms are sorted into separate kingdoms (Plantae and Fungi, respectively).

As a forager, you will want to pay particular attention to plant families. Plants in the same family will share many characteristics, including some rather major ones. For example, if you're allergic to one plant in a family, you will likely be allergic to plants in a shared family. Also, if one plant in a family is toxic, others may be. This doesn't mean that every plant will be, as tomatoes and potatoes are both related to the belladonna (deadly nightshade) plant, but neither are poisonous. Still, it can give you some clues.

You'd be surprised at how many different species of plants there are. For example, when you mention a "dandelion", you could be talking about hundreds of different species.[16] All of these types of dandelions have traits that differentiate them, as well as unique genetic codes. On top of that, you also have to contend with subspecies of plants and cultivars.

- A subspecies of a plant species is isolated from other members of its species and has different physical characteristics.
- A variety of a plant is similar to a subspecies (people use the terms interchangeably) and refers to a naturally occurring variation of a plant within a species.

---

[16] (Prendergast, n.d.)

- A cultivar (or cultivated variety) is a variety of plant species that humans have deliberately cultivated to display specific characteristics.

When foraging, you're likely to come across different subspecies and varieties, but not cultivars (except under unusual circumstances). All of this can seem a bit overwhelming, but it makes more sense in practice. How about another example? The stinging nettle has about six different subspecies.[17] Five of these types of nettles sting, but one (unsurprisingly called the stingless nettle) doesn't. You can find many of these subspecies in different parts of the world. The stingless nettle is only found in Europe, while you can find the European stinging nettle and the American stinging nettle in North America. They have other differences, but this chapter will never end if I go through them all.

Informal Classifications

You can use other methods to classify plants. Formal classifications will only be so helpful when you're foraging; you will primarily use these informal classification methods to identify plants you can forage.

- Latitude. The latitude of the area where you find the plant will give you some clues about what it is. Different plants grow at different latitudes. The broad strokes of these are arctic, temperate, subtropical, and tropical. You won't find a tropical plant in an arctic area, which rules out some options.
- Life cycle.[18] Different plants have different life cycles. These are annuals (the plant completes its life cycle in one year), biennials (the plant takes two years to complete its life cycle), and perennials (the plant has a lifespan of over two years).
- Growing/Flowering season. Does the plant prefer the warm season or the dry season? You could also use the wet vs. dry season, depending on the climate.
- Usage. What is the plant typically used for? Is it eaten as a fruit or vegetable, used for medicinal purposes, as a fiber, as a dye, or has another use?
- Tissue type. Is the plant herbaceous? Or is it a softwood, semi-hardwood, or hardwood bush or tree?

---

[17] (Linnaeus, n.d.)

[18] (*The Classification Of Plants - Annuals, Biennials and Perennials,* n.d.)

- Foliage retention. Different plants react differently to winter. Evergreens retain all their foliage, deciduous plants lose their leaves, and semi-evergreens are in-between.
- Water needs. Different plants thrive in different amounts of water. Desert plants and succulents require far less water than plants found in wet, temperate climates. Some plants (known as halophytes) can tolerate a high salt content in their water, such as a salt marsh.
- Gymnosperm or angiosperm. A gymnosperm ("naked seed") doesn't produce flowers or fruit but releases its seeds in other ways, such as in a cone. Angiosperms produce flowers.
- Monocots or dicots. You can further divide angiosperms (flowering plants) into two categories, monocots, and dicots. The easiest way to tell the difference is to look at the leaves and flowers. Monocots tend to have parallel leaf veins, and flowers have petals in threes. A monocot will never grow true wood or bark. Dicots, however, have branching leaf veins, and the flowers usually have petals in either fours or fives. Dicots are capable (but don't always) of growing wood or bark.

Hopefully, it's starting to come together how you can use the different features of a plant to identify it. If not, don't worry; things will get more apparent as we go on.

Plant Morphology and Classification

First off, we have another definition. Plant morphology refers to the study of the physical form and external structures of a plant.[19] If you remember, we covered the external structures of a plant a few pages ago. Plants are, while perhaps simpler than most animals, fairly complex organisms. However, you can quickly become familiar with the basic structure and organs shared by many plants. The organs of a plant are made up of a limited number of tissue types.

When using morphology, you may look at structures in all kinds of different plants and draw comparisons between them. For example, the leaves of a pine tree and a lettuce plant look very different, but they have the same basic function and similar structures. They're both leaves. You can even look at the spines of a cactus, and you'll find that, yep, cacti spines

[19] (*Plant Morphology*, n.d.) (*Basics of Plant Morphology*, n.d.)

serve the same function as leaves. They look very different at a glance, but developmentally, structurally, and functionally speaking, they're very similar to leaves.

You can also use these comparisons to distinguish between plants and categorize them. Many plants produce flowers, but some don't. However, every wild plant does have a reproductive mechanism. As you inspect each common structure, you may notice other differences. One plant may have a branching fibrous root system, while another may have a taproot (one primary root that digs deep). One plant may form into a bush with many off-shooting stems, while another may grow into a tall plant with a primary stem where all the leaves and flowers will grow. One plant may develop woody tissues and bark, while another will stay green and fleshy.

Another way to use plant morphology is to look at different plant environments and how they impact plant growth and development. Two primary factors impact plant structures:

1. Shared ancestry and genetics
2. Environmental pressures

So, two plants may be similar because they're closely related or because they both grow in the same environment, and their structures have developed to thrive. Two plants with similar genetics might look quite different because they grow in different climates. But you also get plants that have very distant families that develop in similar ways because they have to survive in a particular environment.

Practical Tips for Foragers

Now that you have this botany knowledge under your belt, how can you use it in the field? Here are some practical ways to apply the principles of what we've learned.

Geography

The first way to use your new botany powers is to research your environment. Where are you globally? Some plants only grow in North America, so don't expect to find them in Europe. What is your latitude and climate? What altitude are you? Different altitudes may suit different plants better, and, especially as you reach higher altitudes, you'll find fewer varieties that can thrive.

This will immediately narrow down the types of plants that you should expect to find. If you find a plant that looks edible, but you know that it shouldn't be in that environment, then the chances are that you have a lookalike.

Identification Technique

The best way to identify a plant is by getting up close and personal. No, don't eat it just yet. Use gloves to handle it, as some plants don't react well to being touched. While the root is one identifying feature, it's best to focus on the parts that are easy to see.

Inspect the leaves, flowers, and growth pattern of the stem. First, how big is the plant? How mature is the plant? The level of maturity will help you to identify what you're looking at. Does the plant have flower buds, flowers, or fruits? What color are the leaves and flowers? How are they arranged? How are they shaped? Is the stem fleshy and green or woody and hard? Are there thorns? Are there distinguishing features that are unique to that plant?

The best thing to carry with you is someone who is a field expert of your foraging region. They will be able to teach you on the go. But you should at least have a local field guide when you're exploring the area, as it will inform you of what plants should and shouldn't be there. A field guide will also have pictures of edible plants and dangerous or protected plants. A picture is the easiest way to identify a plant. You can also use mobile phone apps such as Plantsnap[20] to quickly identify a plant.

Make notes in your journal and, if you can, draw a diagram. Compare it to your field guides or phone apps. If the plant looks similar to something that you know is edible, but has some odd differences, then make sure that you figure out what it is before eating it. Yes, I've said this before, and I'll say it again.

I'll demonstrate how useful this identification technique is with an example. Like commonly cultivated carrots, wild carrots (Queen Anne's Lace) are edible and delicious. I once came across a group of plants that looked very similar to wild carrots. They were, however, a good 4 feet taller than your typical wild carrot. Upon closer inspection (and with the help of a handy foraging guide), I noticed that the stems of this plant were hairless and had purple spots. Wild carrots have hairy stems with no purple spots. This plant was no wild carrot, but it was a similar plant known as poison hemlock. There are no prizes for guessing how it got

[20] (Fratt, 2018)

that name. But because I had the right tools and botany knowledge at my disposal, I was able to figure out what plant I was looking at and didn't make any unfortunate mistakes.

Practice these techniques with plants that you're already familiar with. Go out into your backyard or a nearby green spot and try to identify plants using their characteristics.

Plant Families

Once you become familiar with certain plant families[21], then you will always be able to identify them. For example, there are thousands of different species of mint around the world.

So, say you come across an unfamiliar plant. You take a closer look at it and notice several things. First, the plant has square stalks and opposite leaves. When you pick a leaf and crush it, it has an aromatic and slightly spicy smell. You also notice that any flowers are irregular. Congratulations, you've just identified a plant in the mint family.

The Immediate Environment

As well as the geographical location of your foraging site, you should also inspect the plant's immediate surroundings. What's the soil like? Is it dry or wet? Is it rocky, sandy, or full of clay? Is it good quality dark soil full of mulch (rotting leaves)? Is it tightly bound by complex root structures?

You'll notice that certain environments have different types of soil. Beach dunes will likely have sandy soil, while forest floors will probably be covered with mulch. Soil in plains and along hedgerows may be tightly packed and have a high clay content. However, this isn't a hard and fast rule. Different types of soil support different plants.

Another thing you should take note of is the surrounding plants. Many plants grow in the same area, so once you become familiar with certain plants, you can expect to find their companions with them. This could be because both plants adapt to the same environment or benefit from each other.

Before you forage anything, look for any dangers in the environment. Again, I've mentioned this before, but if you see any evidence of nearby industry or unusually sickly plants, then reconsider using this spot to forage.

---

[21] (Vinskofski, 2018)

To Conclude

Understanding basic botany can take you a long way in your foraging efforts. It will provide you with plenty of functional knowledge about the plants you intend to harvest and others that you should avoid. Future chapters will help you expand this knowledge further and apply it to individual plants that are commonly foraged. But before we reach that point, let's explore some simple advice on how you can forage safely and responsibly.

3

Foraging Safely and Responsibly

While I'm trying to encourage you to get out there and go foraging, I can't deny that there are risks involved. According to the American Association of Poison Control Centers (AAPCC), the number of plant and mushroom poisonings among foragers in recent years is on the rise[22]. This is likely, in part, due to an increased current interest in foraging. While I'm all for more people getting involved, the dangers are still evident.

This is why I always put so much emphasis on foraging safely and responsibly. But once you know the dangers and how to avoid them, you can enjoy foraging all the more because you're secure in knowing that you won't get hurt.

The Common Risks Involved in Foraging

I will tackle these risks in more detail, but it seemed prudent to list some of the most common risks involved in foraging. These include:

- Harvesting and eating a poisonous plant due to misidentification
- Exposure to contaminated surroundings
- Aggressive wildlife
- Getting injured while foraging

[22] (Steaven, 2018)

That is by no means an exhaustive list, but these are the dangers most likely to crop up while you're exploring and foraging plants. There are also some minor mishaps that might occur, which you can easily avoid. But let's focus on the most pressing risk involved in foraging, misidentifying plants.

Common Poisonous Plants in North America

There's a common saying among foragers, "anything is edible once". Using the botany skills picked up earlier, you should be able to tell the difference between plants that are edible only once (you know, poisonous) and plants that are safe to eat as many times as you'd like. While I would love to produce an exhaustive list of all of the poisonous plants you can encounter worldwide, it just isn't practical. As I'm native to North America, it makes sense to focus on this area. However, you can find resources online that are specific to your location. Also, you may well encounter some of these plants in other parts of the world. Even if you don't, it's helpful to look at poisonous plant profiles to know what information you need to identify them.[23]

Please see the QR code for photographs of the following plants.

Water Hemlock

Water hemlock, sometimes called "poison parsnip", is commonly mixed up with poison hemlock, despite the two plants being in different families. Both are toxic but have different effects.

Water hemlock looks similar to poison hemlock; it also shares the same forageable lookalikes. The most common of these are wild carrot and yarrow, which is sometimes foraged for medicinal purposes. Water hemlock can be fatal, and it serves as a sobering reminder to only ever ingest something that you're sure of. Even if it looks like an edible plant if in doubt, leave it alone. It's better to miss out on a foraging opportunity than risk poisoning.

- Common Name: Water Hemlock.

[23] (Hodgkins, 2021)

- Scientific Name: *Cicuta Virosa*[24].
- Common Lookalikes: Wild carrot (or Queen Anne's lace) and yarrow. It also looks like poison hemlock, another toxic plant.
- Distinguishing Features: Can reach 8 feet high. Has a hairless, smooth, and hollow stem that stands tall and forms branches. The stem sometimes has purple stripes or a mottled pattern. The roots have thick tubers that smell like parsnip and, when pierced, leak a yellowish liquid that turns reddish-brown. The feathered leaves are alternately spaced in groups that can reach 12-35 inches in length, and each leaf is about 2-4 inches long. The leaf veins end in the notches of the teeth. In spring and early summer, the plant blooms tiny white flowers that cluster in an umbrella shape.
- Habitat: Wet and marshy areas, such as stream banks.
- Geographic Location: Widespread throughout the continental US, also temperate parts of Europe.
- Toxic Parts: Primarily the roots, but the whole plant is poisonous.
- Toxic Dose: Ingesting any amount of poison hemlock can cause poisoning, and even a minimal dose can be fatal.
- Effects On The Body: Poison hemlock is one of the most toxic plants commonly found in North America. When ingested, poison hemlock can cause initial symptoms in as little as fifteen minutes, including nausea, pain, tremors, confusion, and dizziness or weakness. However, the most notable symptom is seizures. These seizures can lead to even more severe symptoms in fatal cases, potentially resulting in coma, respiratory failure, heart failure, and even death. It is possible to die after only a few hours of ingesting this plant.
- How to Treat Poisoning: Activated charcoal if caught quickly, seizure medication, barbiturates, and other supportive medication designed to treat complications.

[24] (Linnaeus, n.d.)

Poison Ivy

This is one of the most common poisonous plants in North America. While poison ivy poison tends not to be fatal, it can ruin an otherwise enjoyable foraging or hiking trip.

- Common Name: Poison ivy. There are two subspecies of poison ivy found in North America[25], they are commonly known as western poison ivy or eastern poison Ivy.
- Scientific Name: *Toxicodendron rydbergii* (western poison ivy)/*Toxicodendron radicans* (eastern poison ivy).
- Distinguishing Features: Grows as small plants, shrubs, or climbing vines. Known by clusters of leaves that alternate on the stem, each with three leaves. This is the origin of a common foraging expression, "leaves of three, let it be". The center leaf is usually the largest. The leaf shape can vary, and the outside of the leaf can be smooth, lobed, or toothed, but they all end at a point. The leaves are reddish in spring, green in summer, and yellow/orange in fall. The poison ivy may bloom in clusters of small, yellow-green flowers from May to July. In the fall, the plant produces tiny white berries.
- Habitat: Woods, along rivers and streams, near lakes, near ocean beaches.
- Geographic Location: Western poison ivy can be found in most states, while eastern poison ivy is primarily found in the eastern states.
- Toxic Parts: Primarily the leaves, but you should avoid touching any part of the plant.
- Toxic Dose: You can get a reaction from merely brushing against poison ivy. Burning poison ivy can release hazardous smoke, which may also cause a reaction.
- Effects On The Body: Poison ivy releases an oil that causes an allergic reaction in humans. It can cause blisters and severe itching for long periods. Some people have no allergic reaction, and extreme cases can cause anaphylaxis.
- How to Treat Poisoning: If you come into contact with poison ivy, wash immediately with either rubbing alcohol or soap and cold water. Wash your clothes. Some lotions or astringents can soothe the itching, and severe cases may call for steroids.

---

[25] (*Poisonous Plants: Geographic Distribution | NIOSH*, n.d.)

Poison Oak

While less common than poison ivy, poison oak can just as easily ruin your trip outdoors. Another similarity to poison ivy is that there are two types of poison oak that you'll find in North America. Despite the name, it isn't closely related to oak trees.

- Common Name: Poison oak. The two species common to North America are Pacific poison oak (or western poison oak) and Atlantic poison oak (or eastern poison oak).
- Scientific Name: *Toxicodendron diversilobum* (Pacific poison oak)/*Toxicodendron pubescens* (Atlantic poison oak).[26]
- Distinguishing Features: It can grow as a shrub in open sunlight and as a long, treelike vine in shaded areas. Poison oak gets its name from its leaves, which usually look like the lobed or scalloped leaves of a true oak. Like poison ivy, poison oak has three leaves with a sizeable middle leaf and two smaller leaves beside it. The leaves are bright green in spring, darker in summer, and red in late summer/fall before falling off in winter. Poison oak has white flowers in the spring and small, round white, or tan globular fruit later on.
- Habitat: Wooded areas, grasslands, varying types of forests.
- Geographic Location: Pacific poison oak is found on the west coast, while Atlantic poison oak is located in the south-eastern states, with Texas as the most western state where you'll find it.
- Toxic Parts: Leaves and twigs.
- Toxic Dose: Even brushing against this plant can cause a reaction. If the plant is burned, the smoke can cause a severe allergic reaction internally and externally.
- Effects On The Body: Poison oak, like poison ivy, releases an oil that causes an allergic reaction for most humans. It causes an itchy rash which can lead to dermatitis and blisters.
- How to Treat Poisoning: Wash affected areas and your clothes. Some lotions can alleviate the symptoms, and severe cases may require steroids.

---

[26] (*Toxicodendron Diversilobum*, n.d.)

Foxglove

Foxgloves[27] are beautiful plants that are sometimes grown for ornamental purposes. Native to Europe, they were brought to the United States for that purpose. Some varieties also grow in the wild.

Foxgloves are also used to create heart medication. However, when ingested, the toxins can be deadly. One common saying is that foxgloves "raise the dead and kill the living". You may find that many plants are sometimes used for medicinal purposes, but that doesn't mean that they're safe to eat. Always do your research first.

- Common Name: Foxglove
- Scientific Name: *Digitalis purpurea*
- Common Lookalikes: Comfrey is a medicinal plant that can be toxic.
- Distinguishing Features: The plant can grow to about 6 feet tall. It has oval-shaped, hairy leaves with a toothed outer edge. The flowers are the most apparent feature. They are usually pink/purple, although they can be white. The tube-shaped flowers grow in a tall shape and hang down. The flowers are visible through summer and early fall.
- Habitat: Woodland edges, along the roadside, hedgerows, and meadows.
- Geographic Location: Continental North America and Europe (mainly the UK)
- Toxic Parts: The entire plant is toxic.
- Toxic Dose: Even touching the plant, let alone ingesting it, can cause poisoning. Tea made from foxglove (usually a case of mistaken identity) can be fatal.
- Effects On The Body: Foxgloves contain a toxin called digoxin, which affects the heart. Early symptoms include vomiting, pain, hallucinations, and headaches. More severe poisonings can cause a slow pulse, heart block, palpitations, seizures, and potentially death.
- How to Treat Poisoning: Severe digoxin toxicity can be treated with an antidote. Other treatments can alleviate the symptoms, including magnesium, lidocaine, and

---

[27] (*Foxglove (Digitalis Purpurea) - British Plants*, n.d.)

atropine. If you suspect that you or someone else has eaten foxgloves, get to a hospital.

Death Camas

This plant is another one with a foreboding name. Death camas are attractive plants that, unfortunately, look very similar to some edible forage. There are several species of plants known as death camas, all of which live in different areas. None of these plants are in the same genus as edible camases. I'll focus on one of the most well-known species, or we'd be here all day.

- Common Name: Death camas or meadow death camas.
- Scientific Name: *Toxicoscordion venenosum* (formerly *Zigadenus venenosus*). It is still often known as *Zigadenus venenosus*, so look out for both names when researching this plant.[28]
- Common Lookalikes: Edible camases (such as the blue camas) and edible onions.
- Distinguishing Features: Onion-like bulbs. Grass-like, rough leaves with parallel veins. The leaves can grow up to 12 inches long and are widest near the stem. The creamy-white flowers have six petals arranged in a star-like pattern. They smell unpleasant but don't smell like onions or other alliums. When it fruits, it forms distinctive cylindrical, papery capsules.
- Habitat: The meadow death camas grow in meadows, forests, and hillsides, near the edible blue camas. You can find other death camas plants in foothills, mountains, rocky areas, savannas, and many other environments.
- Geographic Location: The meadow death camas is primarily found in the western half of North America, including parts of Canada and Mexico. Other species can be found in other parts of North America.
- Toxic Parts: The bulbs and leaves are very poisonous.
- Toxic Dose: Any amount can be toxic, and ingesting only 2-6% of your body weight can be deadly.

---

[28] (*Melanthiaceae – Death Camas – Better Learning Through Botany*, n.d.)

- Effects On The Body: Death camas contain neurotoxins that can cause vomiting, abdominal cramping, low heart rate, low blood pressure, and muscle spasms. It may result in death.
- How to Treat Poisoning: Death camas poisoning doesn't have an antidote. Instead, the symptoms are treated, particularly low heart rate and blood pressure.

How to Test the Edibility of Wild Plants

As a general rule, you should only ever eat something that you know is edible. If you're in any doubt, make some notes and take a picture of the plant. Then you can do some research on it, either using field guides, phone apps, the internet, or, best of all, a more experienced forager friend.

However, while you'll be foraging and bringing things home most of the time, sometimes you're in a different situation. In an emergency survival situation, you might not have the luxury of only foraging the things that you're absolutely sure of. Ideally, you'll be more experienced if you're ever in this situation, but life isn't always ideal. Thankfully, you can spot whether an unfamiliar wild plant is likely to be edible or not[29], without finding out the hard way.

General Tips

These tips will help you to avoid dangerous plants. Some of these tips may also rule out edible plants, but it's better to be safe than sorry. Of course, if you know that a plant is safe, then go ahead.

- Avoid wild plants that look like cultivated ones. For example, cultivated parsnip is a delicious vegetable that you can eat. Wild parsnip is toxic and can irritate your skin.
- Avoid unfamiliar wild mushrooms. Wild mushrooms can be delicious, deadly, or hallucinogenic. I'll revisit mushroom foraging later, but avoid them if you don't know what you're dealing with.
- Don't eat an unfamiliar plant with thorns.
- Don't eat a plant with shiny or waxy leaves.

---

[29] (*Foraging 101: How to Identify Poisonous Plants in the wild (and in Your Garden)*, n.d.)

- Avoid plants with umbrella-shaped flowers.
- Avoid white, yellow, or otherwise brightly colored unfamiliar berries.
- If the plant has a milky or discolored sap, don't eat it. Some of these plants, like dandelions, are safe to eat but steer clear if you're unsure.
- Don't eat anything that smells like almonds, including wild almonds or the pits of stone fruits. While you're at it, avoid the seeds and leaves of apples and stone fruit trees. They contain something called amygdalin, which turns into cyanide when eaten.
- Plants with beans, bulbs, or seed-bearing pods are often poisonous.
- Plants with woody stems or leaves are more likely to be toxic.
- "Leaves of three, leave it be" - these plants may be poison ivy or a similar species.
- Don't burn any plants that may be poisonous. Sometimes the smoke is dangerous.

If you follow these general rules and pay attention to the warning signs[30], then you're less likely to eat any poisonous plants accidentally. This doesn't mean that you can stop being cautious. There's still another step to take before you should eat a wild plant.

The Universal Edibility Test

Remember that you should always be sure that you've correctly identified a plant before consuming it. Poisonous plants are a serious matter, as I hope I've established. The universal edibility test[31] is only meant to be used in emergency situations. Suppose you need to survive off the land and have no other food supply (such as wild plants that you can reliably identify). In that case, the universal edibility test is a valuable skill to help you to determine if a plant is safe to eat.

With all that in mind, what is the universal edibility test?

The universal edibility test is a method where someone can determine whether or not a plant is edible. It isn't completely foolproof, but it will give you a good shot of eliminating any toxic plants. The test involves breaking down a plant into its major parts and testing them

---

[30] (Bryant, n.d.)

[31] (Licavoli, 2021)

individually over 24 hours. You should also focus on a plant that's growing in abundance. There's no point taking 24 hours to establish that a single plant part is edible if barely any of it grows in the area.

The best way to show you how it works is to provide a step-by-step process.

1. Fast

Ideally, you should have fasted for at least 8 hours before beginning this test. This is so that you can be sure that any reaction is from the plant you're testing, not something else that you've eaten.

If you're in a survival situation where you've run out of food, which is the most likely reason for trying this test, this shouldn't be too implausible.

However, you can finish steps 1-3 before fasting after step 4.

2. Divide the Plant

Some plants have both edible and toxic parts. This means that you'll want to test each part separately. Divide the plant into its primary parts and separate them. If you can, use gloves. If the plant causes an immediate reaction when your skin comes into contact with it, stop.

- Peel off the leaves
- Pull out any fruit
- Remove any seeds from fruit or flowers
- Pull off and collect flowers and buds
- Break up the roots
- Lay out every plant part, including the stem, in an organized manner

Now you want to inspect every part and study for warning signs. If you see worms, parasites, or any other signs of rotting, then discard the plant parts and start again with a different sample. Any brightly colored or otherwise discolored plants are more likely to be toxic, so avoid those. If the plant weeps a milky or yellowy sap, then it likely isn't safe.

3. Smell

The smell of a plant isn't a reliable way to tell whether or not it's edible, but certain smells can warn you to stay away from the plant.

- Acidic odors
- Strong, foul smell
- Musty or moldy smell
- Pears or bitter almond smell (possible sign of cyanide)

4. Rub

This part of the test will indicate whether your skin reacts to the plant or not. You can do this part before the fasting phase.

Select the plant part that you think is most likely to be edible. Crush it and rub the juice and crushed plant part on your forearm.

Your skin may not react immediately, so you will have to wait for 8 hours before moving on to the next step. You can drink water during these 8 hours but don't eat anything.

If you notice any soreness, itching, swelling, blistering, or any other signs of contact dermatitis on your skin, then discard the plant and wash your hands (if possible). Some creams can relieve the symptoms, as can pressing a cold rag against the area.

If 8 hours pass and you have no symptoms, you can move on.

5. Cook

Cooking wild plants is the safest way to eat them. Cooked plants are easier on your digestive system, and cooking can make some otherwise poisonous plants safe to eat (like elderberries). Cooking also gets rid of any germs or other unwelcome visitors.

Cook a tiny portion of your chosen plant part in the way that you'd prefer to eat it. If you don't have any way to cook it, you'll have to eat it raw.

6. Taste

Once it's cooked (or not), take the plant part and hold it to your lip for 3 minutes. If you feel a burning or tingling sensation, then discard the plant part and start again with a different one.

If there's no reaction, place the plant part on your tongue and leave it in your mouth for at least 15 minutes. Resist the urge to chew or swallow it; just leave it alone.

If you have an adverse reaction, such as burning or tingling, then spit it out and wash your mouth out with water. You should also spit it out if it tastes soapy or overly bitter.

Remember that it doesn't have to taste particularly nice to be edible. You're only looking for a reaction that indicates toxicity.

7. Chew

Unfortunately, you can't eat the plant yet. The next step is to thoroughly chew the plant to crush it and then hold it in your mouth without swallowing for another 15 minutes.

You're looking for similar signs as before, such as burning, tingling, numbness, or a soapy/bitter taste. If any crop up, spit it out and rinse your mouth with water.

8. Swallow

After the plant part has sat in your mouth for 30 minutes without any reactions, you can finally swallow it.

Once you've swallowed the plant part, you have to fast for another 8 hours. You can drink water, but don't eat any more samples of that plant part or any other food. It's essential to keep the test to a small amount so that, if it is toxic, you only get a minimal dose. The 8 hours will give you time to fully digest the plant part so that you can determine if it has any effect on your body.

If you feel nauseous or have any symptoms of poisoning, try to induce vomiting and flush your body by drinking a lot of water. Wash your hands as well.

9. Repeat

If you don't have any side effects after 8 hours, prepare the plant part in the same way you had before and eat about a palmful (or ¼ of a cup) of it. Wait another 8 hours to see how you react to the larger portion of the plant.

If you have poisoning symptoms, then follow the same procedure as before. If you don't, then congratulations! That plant part is probably safe to eat.

In any case, you will have to wait at least 8 hours before starting the test again with a different plant part or a new plant.

As you can imagine, this takes a lot of time. However, you don't have to sit around waiting while you fast. If you're in a survival situation, then you should multi-task. This test might seem a bit extreme, especially if you have no food with you and you're hungry. But remember that even if a plant "only" causes gastrointestinal symptoms, vomiting and diarrhea can quickly become very serious in a survival situation.

Common Signs of Poisoning

One of the major parts of performing the universal edibility test is looking out for potential poisoning symptoms. Poisonous plants can affect different parts of your body, including your digestive system, circulatory system, nervous system, heart, lungs, and skin. Here are some of the most common indicators that you've eaten or encountered something toxic:

- Rash or blisters
- Nausea, vomiting, abdominal pain, diarrhea, or other digestive problems
- Burning or irritation on the lips or within the mouth
- Dry mouth or excessive salivation
- Sweating
- Dizziness
- Muscle weakness
- Difficulty breathing
- Drop in blood pressure and breathing rate
- Unusually quick or slow heartbeat

Severe symptoms of poisoning may include:

- Seizures
- Paralysis
- Collapsing
- Severe disorientation
- Unresponsive to touch or sound
- Vomiting blood

If you think you've been poisoned, seek medical attention as soon as possible. The sooner you can get it treated, the better. Ideally, you should find out what plant has caused the poisoning. Either find out the name or, if possible, gather a sample of it. This will help medical professionals to find the best treatment.

Safe and Responsible Foraging

When foraging, your primary safety concern should be looking out for poisonous plants. However, there are other aspects of safe and responsible foraging you should know about.[32] First, let's tackle where you forage. Here are some dos and don'ts for foraging locations.

- Do ask permission if foraging on private land. Don't worry; they'll probably say yes.
- Do research the legality of foraging on public land. Some governmentally owned land allows foraging, while other municipalities prohibit it. Check the rules of the area first and follow the legal guidelines.
- Do research local plant and animal life and take note of any dangers. Don't forage in and around otherwise contaminated areas.
- Don't forage near industrial areas that might be affected by water run-off.

Once you've found the right location and you're sure that it is safe and legal to forage there, you can move on to making sure that you forage safely and responsibly. Here are some more dos and don'ts to follow while you're out there.

- Do wear appropriate clothing and shoes.

---

[32] (Schipani, 2019) (Kubala, 2021)

- Do bring appropriate gear, including a field guide, food, water, cutting/digging tools, bags/baskets, and other equipment to help you forage.
- Do follow any area regulations, follow trails if necessary, and avoid off-limit spaces.
- Do respect the area and fellow foragers or hikers; it never hurts to be polite, and they might point out a good foraging spot nearby.
- Do take care when foraging from plants with defense mechanisms. I learned this one the hard way, falling into a blackberry bush isn't worth getting at the juicy berries in the middle.
- Don't trample on delicate plants or environments, such as streamsides.
- Don't litter or leave the place a mess.
- Don't approach wild animals.
- Don't hog resources; leave something for future foragers and wildlife.
- Don't destroy or dig up whole plants; let them grow back in the future.

To Conclude

Foraging can be a rewarding and enjoyable experience as long as you keep in mind safety precautions and make sure that you forage responsibly. If you start good habits early, they'll become second nature. Now that we've learned about what you shouldn't forage let's move on to some common plants that are safe to eat.

4

Profiles of Edible Wild Plants, Part 1

There are an estimated 120,000 varieties of edible plants out there for us to forage (some put the figure even higher). This means that we're spoiled for choice, but it's impossible for anyone to be able to distinguish between every variety. The best approach, especially for a beginner forager, is to become competent at identifying the edible plants that you're most likely to come across in your foraging areas. Use your botany knowledge to help you to differentiate them from toxic lookalikes. Once you have some workhorse plants that you can reliably spot, you can start to look into more uncommon varieties.

As with the poisonous plants, these plants all grow in North America. Many of them also grow in other parts of the world. Without any further ado, let's look at some of these common edible plants.[33]

Please see the QR code for photographs of the following plants.

Burdock

Burdock is a common plant used in several food and drink recipes worldwide. In the UK, for example, it's part of a drink that tastes similar to root beer. Burdock has a distinctive and earthy flavor, which some people enjoy and others dislike. The only way to find out if you like it is to try it yourself.

---

[33] (Coelho, n.d.), (*40 Most Common Edible Wild Plants in North America*, 2015), (*Edible Plants Guide*, n.d.), (Orr, n.d.), (*Wild Edible Plants of the Pacific Northwest*, n.d.)

- Common Name: Burdock.
- Scientific Name: *Arctium lappa.*
- Edible Part/s: The taproot, the leaves, and the pith of the stem. Older burdock leaves are tough, so they taste better when cooked. You should also cook the other edible parts before eating.
- Habitat: Riverbanks and rich, loamy soil.
- Geographic Location: Grows throughout North America, Europe, Asia, and Australia.
- When to Forage: Burdock is biennial, and it's best to forage it at any point during its second year of growth. It's easy to tell when it's in this stage.
- Distinguishing Features: During the first year of its growth, burdock presents as a rosette of green leaves at ground level. It grows into a 5-foot tall, bushy plant with large, long leaves during its second year. The lower leaves have wavy edges. The leaves are dark green above and white or wooly underneath. The most notable features are the irritating burrs that give it its name, which are green, spiky globes. From June to October, Burdock will produce pink or purple flowers.
- Lookalikes: Burdock looks a lot like cocklebur. Cocklebur leaves are toxic when raw, so they need to be cooked.

Bamboo

Bamboo is technically a grass and famously grows incredibly quickly. There are hundreds of bamboo species, and while they're most well known for growing in East Asia, some grow in other parts of the world, like North America. Bamboo can crowd out other plants and become an invasive monoculture (an area with only one species). Not every bamboo is edible, so make sure you know what type you're eating.

In North America, one of the most widespread edible bamboo species is golden bamboo (*Phyllostachys aurea*). But you may find plenty of other species. Depending on the type of bamboo, the shoots might taste sweet or savory. Thankfully, bamboo is very distinctive. Bamboo is also a good foraging option for those who like crafting things. You can even use the non-edible varieties to make furniture, beanpoles, rafts, baskets, and other things. This makes it a fantastic foraging option for the survivalists among us.

- Common Name: Bamboo.
- Scientific Name: *Bambusoideae*. in North America[34], you will find bamboo with the tribes[35] *Arundinaria* or *Olmeca.*
- Edible Part/s: The young shoots. Boil before consuming.
- Habitat: Depending on the species, prefers either temperate (*Arundinaria)* or tropical (*Olmeca*) areas. The most common North American species prefer wet soil and are surprisingly cold tolerant. They are often found near rivers and streams and grow in concentrated thickets, leaving little space for other plants.
- Geographic Location: You can typically find North American *Arundinaria* species in Florida and the Deep South, as well as some other south-eastern states. *You can find Olmeca* species in Mexico. Other bamboo species can be found in South America, Asia, Africa, and Oceania. Bamboo typically doesn't cope with cold or frost.
- When to Forage: Spring is your best bet, but it can depend on the type of bamboo.[36]
- Distinguishing Features: Bamboo is a tall, edible plant with a distinctive ribbed cane that can grow up to 100 feet high. Most edible varieties are smaller, reaching up to 26 feet high, depending on the species. As well as the distinctive canes, bamboo grows branches with long, thick leaves.
- Lookalikes: Some bamboo species aren't edible, but these species are uncommon, and boiling the shoots in several changes of water usually eliminates any toxic components.

Blueberry

Blueberries are sweet, slightly tart, and generally delicious. They can be eaten raw or used in a variety of recipes. Blueberries are famously very healthy, so they make a fantastic addition to your diet.

---

[34] (*Genus Arundinaria: Native Bamboo of North America,* 2020)

[35] In taxonomic classification, a tribe is larger than a genus but smaller than a family.

[36] (Allday, n.d.)

Several species of blueberries grow in North America. The species that we're most familiar with is *Vaccinium corymbosum*, or "highbush blueberries". This is the type of blueberry cultivated and sold in the United States. In the North American wilderness, you may come across this species or any of the other dozen or so species.

- Common Name: Blueberry.
- Scientific Name: *Vaccinium Cyanococcus.*[37]
- Edible Part/s: The berries are edible and delicious. The leaves are also edible and can be used to make tea. Blueberry leaves contain even more antioxidants and nutrients than the berries themselves.[38]
- Habitat: Sunny areas close to the water. Blueberries prefer acidic soil.
- Geographic Location: Most blueberries grow wild in Canada, the northern states, and the eastern areas of North America. Some species, such as the rabbiteye blueberry, can be found further south. You can also find blueberries in Europe.
- When to Forage: The peak season for foraging blueberries is late summer, but they can fruit from May to September.
- Distinguishing Features: Blueberry plants grow as bushes or shrubs and often grow pretty tall. The stems have no thorns. The leaves can be either evergreen or deciduous and vary in size and shape. Blueberry flowers are bell-shaped and can be white, pink, or red. The berries are small and have a flared crown on one end. They grow in clusters, not as single berries. They ripen from a pale green color to a uniform blue. When cut open, blueberries have green flesh.
- Lookalikes: Edible huckleberries, whortleberries, and bilberries (common in Europe) look similar to blueberries but have red or purple flesh instead of green. Unfortunately, some toxic berries[39] can be mistaken for blueberries. The moonseed berry is dark blue and grows in clusters, but it has one single moon-shaped seed and tastes bitter. Pokeweed grows dark purple berries that look similar to blueberries, but

---

[37] Cyanococcus is a section of the Vaccinium genus. All blueberry species are sorted into the Cyanococcus section.

[38](*Health Benefits of Blueberry Tea*, 2020)

[39](*Berries That Looks Like Blueberries*, n.d.)

the color is slightly off, and they grow in distinctive cylindrical clusters on a light purple stem. Deadly nightshade berries are very toxic. They're usually black and don't have a flared crown.

Prickly Pear Cactus

There are a few different cacti known as the prickly pear cactus, and they're all in the *Opuntia* genus. I'll be mainly talking about the most widespread species, the *ficus-indica*, but every prickly pear is edible.

When foraging for the prickly pear cactus, watch out for the long spines and the even more annoying hairs (called glochids). The spines will hurt, but the tiny hairs will get stuck in your skin. Wear long, thick clothes when harvesting this plant, including gloves. Some people use tongs to pick the fruit; then, they can burn off the spines and hairs with a torch.

- Common Name: Prickly pear cactus.
- Scientific Name: *Opuntia ficus-indica.*
- Edible Part/s: The flesh and fruit are edible; just watch out for spines!
- Habitat: Dry, open areas such as prairies and foothills.
- Geographic Location: It is prevalent in Mexico, but you can find them in any warm and arid or semi-arid area of North America. It's native to Western and South Central United States and the sandy coastal beaches along the East Coast.
- When to Forage: The fruit should be harvested in August, when it's turned deep red.[40] The pads can be foraged year-round.
- Distinguishing Features: These many-branched cacti have broad, rounded, flat pads. The pads are adorned with long spines. They flower and fruit along the top edge of the pads. The round fruits are deep red and covered with large spines and delicate, almost invisible hairs that can irritate the skin.
- Watch Out For: The spines can hurt, but the fine glochids can cause even more trouble if they get stuck in your skin and you don't deal with them. Protect your skin and use

[40] (Grant, 2020)

glue or tape to remove some of the hairs. If they aren't removed, you may need medical attention.

- Lookalikes: Other types of cacti are poisonous, so make sure that you're foraging the right kind of cacti.

Acorn

When you forage acorns, you should watch out for any signs of spoilage. Break open a few while you're foraging, and look out for black, moldy, or half-full acorns. These are bad. The meat should be consistently orange or yellow. Discard any with defects, like holes that indicate grubs. The acorn should be firm, so discard any soft ones. When you get back home, put your collection in water and only keep the ones that sink.

You can dry acorns, roast them, or even crush them to create a flour substitute. They can be sweet, nutty, and oily, but some may be bitter. Acorns provide starch, which is hard to find in the wild. This makes them an excellent option for survivalists or anyone who wants to challenge themselves to be more self-sufficient.

- Common Name: Acorn or oak tree.
- Scientific Name: *Quercus.*
- Edible Part/s: The nuts are edible.
- Habitat: You can find oak trees in hardwood forests and fields. Different species have a wide range of habitats.
- Geographic Location: Native to most of the Northern Hemisphere. North America has a vast amount of oak diversity.
- When to Forage: Fall, namely September to November.
- Distinguishing Features: Acorns are easily identifiable. Oak trees are large and usually have lobed deciduous leaves (although coniferous species also exist, as do some with different leaf shapes). Acorns generally look the same; the brown nuts are ovoid and about an inch long. They're most identifiable by their little green "caps", which often have a short stem attached.
- Watch Out For: Acorns are easy to identify, but you ideally want acorns from the white oak group, as they have fewer tannins. The tannins make the acorn more bitter.

White oak trees usually have rounded, lobed leaves and peelable bark. Red oak trees (which have more tannin) have spikier leaves and tight, dense bark.

Bastard Cabbage

Bastard cabbage, otherwise known as common giant mustard or turnip weed, is an invasive plant native to Eurasia. It was introduced to Texas and has a nasty habit of creating monocultures, pushing out native plants in the process. Invasive plants are an ideal foraging option because you're killing two birds with one stone. You're getting rid of a damaging and invasive species and getting a meal out of it.

Bastard cabbage is easy to identify, and while you can eat it raw, the leaves taste best cooked.[41] Cooking them will eliminate the offputting hairy texture and the bitter taste. You'll be left with a subtle spinach-like flavor.

- Common Name: Bastard cabbage or common giant mustard.
- Scientific Name: *Rapistrum rugosum.*
- Edible Part/s: Leaves, stems, seedpods, and flowers.
- Habitat: Mainly ditches and fields. Likes recently disturbed areas but can be found near streams and forests.
- Geographic Location: Native to Eurasia. Very common in Texas but has spread to the West Coast, as well as some eastern states and eastern Canada.
- When to Forage: Like other mustards, bastard cabbage prefers cool weather, and it's best to harvest them from fall through spring.
- Distinguishing Features: The bastard cabbage has a rosette of large, broad, and deeply lobed leaves at the base of the stem. The leaves further up the stem are narrower and less dramatically lobed. The flowers grow in bunches off the stem and have four yellow petals that form an X.
- Watch Out For: Depending on where you forage, someone may have sprayed the plants with weed killer.

[41] (*Eating Invasive Bastard Cabbage for the First Time*, 2016)

- Lookalikes: Mustard plants all tend to have the same distinct flower shape, making them easy to identify. If you mix it up with another mustard plant, don't worry, they're all edible, and most of them have a pleasant peppery taste.

Bull Nettle

Like many nettles, bull nettles are edible and surprisingly delicious. However, they have the same downsides as other stinging nettles. Bull nettle has a particularly nasty sting that can cause intense pain and stinging. Even worse, it lasts for hours. When foraging, wear protective gear and consider using tongs or pliers to avoid the stinging hairs.

So, why go to the effort of foraging for bull nettle? Simply put, the seeds are delicious. Let the seed pods dry before removing the large, kidney bean-shaped seeds. You can also eat the long taproot. Peel the root, then either boil or roast it to eat it. Just get rid of the fibrous center of the root, as it's too tough to eat.

- Common Name: Bull nettle or Texas bull nettle.
- Scientific Name: *Cnidoscolus texanus.*
- Edible Part/s: The seeds and the taproot
- Habitat: Tolerates arid areas and prefers sandy, well-drained soil.
- Geographic Location: Texas and Southeastern United States.
- When to Forage: Summer and fall.
- Distinguishing Features: The bull nettle grows to about 2 feet high. It has an erect stem that turns woody with age. When it blooms, the flowers are white with five petals each. The leaves have multiple deep lobes. The most noticeable feature is the long, stinging hairs covering the whole plant. The green seedpods have particularly long stinging hairs.
- Watch Out For: Bull nettle, like other stinging nettles, have fine hairs that can cause irritation. In rare cases, this irritation can become severe enough to require treatment. As well as the hairs, the plant has very acidic secretions that irritate the skin. Remove the hairs with tape and apply a baking soda paste to neutralize the acid if stung.

Asparagus

You're likely already familiar with cultivated asparagus, as it's easy to buy from the grocery store. But, as it turns out, you don't need to grow it. Asparagus grows wild in North America and Europe.

You can cook it in much the same way as store-bought asparagus. Simply adjust your favorite recipe.

- Common Name: Asparagus.
- Scientific Name: *Asparagus Officinalis.*
- Edible Part/s: The young shoots are edible.
- Habitat: Can be found in fields, prefers rural areas where there is nearby water. Often located near roadsides. It prefers full sun, so you won't find it in shady areas.
- Geographic Location: All over North America, as well as most of Europe and western Asia.
- When to Forage: Early spring, when the shoots are young and haven't gone woody yet.
- Distinguishing Features: Asparagus grows straight, stout stems with distinctive scale-like leaves. It also has feathery foliage and white or yellow bell-shaped flowers. These flowers eventually form a red berry. When foraging asparagus, you will be looking for young shoots. Ideally, the shoots won't be woody yet, and the plant won't have formed flower buds.
- Watch Out For: If you're foraging near a roadside, ensure that the nearby water source is safe and that someone hasn't sprayed the plant with weedkiller.
- Lookalikes: Young wild indigo (*Baptisia*) looks like a type of purple asparagus in early spring. Wild indigo shoots are edible in small quantities but poisonous if you have too many of them. If you aren't sure, take the shoots home and put them in the fridge. Asparagus will stay firm, while wild indigo will go limp very quickly.[42]

---

[42] (*Baptisia, a Wild Asparagus Look Alike*, n.d.)

5

Profiles of Edible Wild Plants, Part 2

Once you've become accustomed to a few edible plants in your local area and know how to identify them reliably, it's time to expand your list of workhouse forage. You'll also likely start spotting wild edible plants everywhere as you realize how common they are. Many wild edible plants are either unwanted weeds or invasive species, so it's easy to build up a habit of collecting a few wherever you go. There's something about eating off the land that you just can't beat.

Please see the QR code for photographs of the following plants.

Daisy

There are a lot of different species of daisies, but most people are familiar with the common daisy, otherwise known as the English daisy or the lawn daisy. As you'd expect, this daisy is native to the United Kingdom and much of Europe but has since spread to North America.

The great thing about the daisy is that it's incredibly common, easy to identify, and edible. You can eat daisies raw or cook the leaves. They make an excellent addition to sandwiches or salads, or you can snack on the go. Other species of daisies are edible as well, but some are bitter and less palatable.

- Common Name: Common daisy.
- Scientific Name: *Bellis perennis.*

- Edible Part/s: The flowers, stems, and leaves are all edible.
- Habitat: Meadows and grasslands.
- Geographic Location: All over North America, common in Europe.
- When to Forage: Blooms in spring and summer. Depending on your climate, you can find them from March to early October.
- Distinguishing Features: You can easily spot a daisy by the small flowers it produces. The petals are typically white, although they're sometimes tinged with pink or purple. Some daisies have double flowers. The petals surround a distinctive yellow center. Daisies are short flowers with a single stem and rounded leaves that grow near the ground.
- Watch Out For: If you're foraging in disturbed areas (like roadsides and places near construction sites), look out for herbicides. Plants that have been sprayed look obviously unhealthy, but it helps to ask the landowner if possible.
- Lookalikes: Some daisy species aren't as palatable. For example, the chamomile daisy has bitter greens.[43] Some people can get a reaction when touching the chamomile daisy, so eating it isn't the best idea if you're sensitive to it. You can identify chamomile daisies by their tall, sturdy stems and frilly leaves.

Elderberry

Elderberries are fantastic forage because they're so versatile. You can use the pleasantly tart elderberries to make pies, jams, jellies, and even wine—the berries pair exceptionally well with blackberries, another fall forage. The flowers are also nice to add to eat raw or to use in white wine. Elderberries are also very good for you.

However, the rest of the shrub is toxic, and you shouldn't eat any of it. While you can have a few ripe berries raw, they still contain the same toxic components as the rest of the plant, which can cause vomiting and diarrhea. Eating too many raw elderberries is one of those mistakes you only make once, as the results are too unpleasant to repeat. They taste better when you cook them anyway.

---

[43] (Davis, 2015)

- Common Name: Elderberry or elder bush.
- Scientific Name: *Sambucus canadensis*.
- Edible Part/s: Flowers and berries.
- Habitat: The edges of woods.
- Geographic Location: Throughout eastern North America. Some elderberry species grow in other parts of North America, as well as Europe, South America, Australia, and Asia.
- When to Forage: The flowers can be foraged in summer, while the berries are ripe in fall.
- Distinguishing Features: The elderberry is a shrub that can grow to over 9 feet tall. The leaves are arranged in opposite pairs and grow in groups of 5 to 9. Each leaf is about 4 inches long and has very closely spaced teeth. It's easier to look for the flowers or berries. The flowers are small and white and grow in large clusters. The small, bb-sized elderberries start white, then ripen to a dark purple or black. The berries also grow in large groups that droop on the plant.
- Watch Out For: Only the flowers and berries are edible; everything else is toxic. You should cook the berries before eating.
- Lookalikes: There are a few elderberry lookalikes to watch out for. Pokeberry (or pokeweed) has similar berries, but the fruit hangs in small rings rather than large clusters. Dwarf elderberries are also similar, but the berries stand up high rather than drooping down. Both of these lookalikes have berries that you should cook before eating. Another lookalike is Hercules' club, which has similar-looking, but poisonous berries. However, Hercules' club has thorns. The one to really watch out for is water hemlock, which can be lethal. Water hemlock has much narrower leaves with more widely spaced teeth. If you look closely, you'll notice that the leaf veins end in a notch of the teeth.

Hazelnut

All hazels are edible and have similar features, so once you get the hang of spotting them, you can harvest nuts to your heart's content. Wild hazelnuts tend to be smaller than commercially grown nuts, which are otherwise known as filberts.

Once you harvest the hazelnuts, let them dry for a couple of weeks before removing the bitter brown husks. Some hazelnuts have spiny husks (or involucres), so gloves might be a good idea. Remove the husks (wear gloves if they're spiny), and you're left with some perfect little hazelnuts. They'll keep inside their shells for a long time; crack the shells before eating raw or roasting.

- Common Name: Hazelnuts or American Hazel.
- Scientific Name: *Corylus americana.*
- Edible Part/s: The nut is edible, either raw or roasted.
- Habitat: Sunny areas along streams, fields, and near forests. They prefer rich, moist soil.
- Geographic Location: Eastern and central United States and southern parts of eastern and central Canada. You can find other species of hazel in western North America, as well as Europe and other parts of the world.
- When to Forage: Between July and October
- Distinguishing Features: Hazel trees aren't giant, only reaching about 12 feet tall. They sometimes look more like big shrubs. The trees have heart-shaped leaves that are alternately arranged on the branches. The leaves are usually about 6 inches long and 4½ inches wide. Each leaf is darker on the front than the back and has doubly serrated edges. In spring, the tree will produce catkins that hang down. The nuts grow enclosed in 2 distinctive leaf-like bracts.[44]
- Watch Out For: Most nuts are foraged in fall, but hazelnuts are usually ready earlier. If you look later in the season, something might have beaten you to the punch. Avoid hazelnuts infested with insects.

---

[44] A bract is a modified leaf structure that typically grows beneath a flower or fruit. (*Bract | Plant Structure | Britannica*, n.d.)

Japanese Knotweed

Japanese knotweed is an infamously invasive species, to the point that it's illegal to propagate it deliberately. Japanese knotweed can damage roads, foundations, and buildings as it grows. It's also tough to get rid of once it's established, requiring constant upkeep and repeated weedkiller treatments.

The good news is that Japanese knotweed is an excellent source of food. For the best results, harvest young shoots that are about ½ a foot tall for the best results, although you can eat those under a foot tall as long as you remove the outer bark. Boil the shoots before eating. It tastes similar to rhubarb.

- Common Name: Japanese knotweed.
- Scientific Name: *Fallopia japonica.*
- Edible Part/s: Young shoots (under a foot tall, ideally ½ a foot).
- Habitat: Can thrive in most temperate environments, often found near roadsides and wasteland.
- Geographic Location: Throughout North America, also found in Europe and Asia. Japanese knotweed is, unsurprisingly, native to East Asia.
- When to Forage: Early spring, usually in April and May.
- Distinguishing Features: Mature knotweed grows up to 10 feet tall and forms dense thickets. It looks a bit like bamboo but has broader leaves that grow in an alternating zigzag pattern. The mature stems are hollow. Japanese knotweed flowers bloom in late summer. They are small and cream in color and grow in clusters called racemes.[45] The young sprouts are green with red or purple mottling. They aren't hollow.
- Watch Out For: Only the young sprouts are edible, so don't eat an older plant. Try to avoid trampling through the knotweed, so you don't accidentally spread it any further. Ensure the Japanese knotweed isn't treated with herbicides when you harvest it.
- Lookalikes: There are quite a few lookalikes. Some are edible, while others, such as bindweed, are toxic.

---

[45] A raceme is a group of flowers that grow on a short stalk.

Lamb's Quarters

In some parts of the world, lamb's quarters is deliberately cultivated as a food crop. However, it's considered a weed in North America, especially in crop fields.

You can eat lamb's quarters raw, but it's best to cook it if you plan to eat a lot as it contains a lot of oxalic acid, which can damage your kidneys. You can cook the shoots and leaves and eat them like greens. It has a mild, sweet, and salty taste and is an excellent source of vitamins A and C.

- Common Name: Lamb's quarters or white goosefoot.
- Scientific Name: *Chenopodium album.*
- Edible Part/s: The leaves, shoots, flowers, and buds are all edible. You can sometimes harvest the seeds, but this can be tricky.
- Habitat: Can grow everywhere, especially in wasteland.
- Geographic Location: Throughout North America. It's prevalent in Europe, and you can find it on every continent except Antarctica.
- When to Forage: Ideally, forage this plant in spring, when it's still young.[46] This way, the leaves are more tender and have less oxalic acid. They're also sweeter and more palatable.
- Distinguishing Features: Lamb's quarters typically grows upright, reaching around 4 feet high. It has several different shapes of leaves. The lower leaves are toothed and diamond-shaped, while the upper leaves near the flowering stems are long and narrow. The leaves also tend to have a white powder underneath, which is what makes them slightly salty. The small, green flowers grow in clusters on the spiky stems.
- Watch Out For: This plant contains oxalic acid. You can get rid of it by cooking it in water and discarding the water, but it shouldn't be a problem if you only eat a small amount.

[46] (*How to Use Lambsquarter From Root to Plant to Seed,* n.d.)

Dandelion

Dandelions are one of those weeds that everyone is familiar with because they're simply everywhere. There are loads of species of dandelions about, but the most common species is known, rather boringly, as the common dandelion.

Dandelions are a forager's dream because they're everywhere. They're also straightforward to spot and harvest. You can eat the flowers and leaves raw as salad greens (they're similar to mustard greens), although older leaves are less palatable and better when cooked. You can use the flowers to make wine or jam. Even the taproot can be foraged, cooked, and eaten, or used to make a beverage (similar to burdock).

- Common Name: Dandelions or common dandelions.
- Scientific Name: *Taraxacum officinale.*
- Edible Part/s: Flowers, flower buds, leaves, and taproot.
- Habitat: Lawns, disturbed land, roadsides, or anywhere with moist soil.
- Geographic Location: You can find the common dandelion throughout North America and Europe. Other species grow elsewhere.
- When to Forage: Dandelions bloom from May to October, but it's best to forage them from May to June, which is when they're more likely to have flowers and younger leaves.[47] You can harvest the root at any time of the year.
- Distinguishing Features: You can usually spot dandelions by their distinctive bright yellow flower heads that turn into tufty silver seed bulbs (known as clocks or blowballs) that blow away in the wind. The leaves are long with deep lobes that reach almost the main leaf vein. The stem has a milky sap when it's broken. Dandelions also have an extensive taproot, which is part of what makes them difficult to eradicate.
- Lookalikes: A few plants known as false dandelions look a bit like dandelions. However, most of these are edible, such as cat's ears and hawkweeds.

---

[47] (*Dandelion | Garden Organic,* 2022)

Garlic Mustard

Garlic mustard is yet another invasive species that makes for a fantastic foraging target. Even better, it's delicious. The same things that humans find so delicious are unpalatable to deer, which is why it thrives so much in North America.

The leaves, stems, and flowers have a garlicky taste with a hint of mustard and work great in salads, as a flavoring, or to make pesto. The roots smell and taste a bit like horseradish, so you can use them as a seasoning. You can use the seeds as a seasoning or to make a kind of mustard.[48]

- Common Name: Garlic mustard.
- Scientific Name: *Alliaria petiolata.*
- Edible Part/s: The whole plant is edible.
- Habitat: Can be found in shady forests.
- Geographic Location: Much of North America, except the southern border states and Nevada, Wyoming, South Dakota, and Wyoming. Native to Europe and very common in Britain.
- When to Forage: Forage the leaves in spring and the seeds in summer.
- Distinguishing Features: In the first year of growth, garlic mustard will appear as a rosette of round leaves. In the second year, you will spot tall flowering stems with small white flowers and triangular or heart-shaped leaves. The easiest way to identify garlic mustard is by smell; the leaves smell of garlic when crushed.
- Lookalikes: There are several lookalikes, including ground ivy and hairy bittercress, but you can tell the difference by crushing the leaves and checking for a garlicky smell.

Watercress

You commonly find watercress in the grocery store, but you can find it in the wild as well. Use scissors to snip some cress to take home and use it quickly, as it can wilt very soon after being picked.

---

[48] (Docio, n.d.)

Rinse thoroughly and enjoy raw, either in salads, in sandwiches, or as a garnish. You can cook watercress, but it doesn't taste as good.

- Common Name: Watercress.
- Scientific Name: *Nasturtium officinale.*
- Edible Part/s: Sprouts, leaves, flowers, and tender stems.
- Habitat: Grows in shallow running water. It prefers slightly alkaline water.
- Geographic Location: Found throughout the world.
- When to Forage: Best harvested in the winter or spring, before it flowers. After flowering, the watercress will be bitter but still edible.
- Distinguishing Features: It grows in a dense mass. It can grow a couple of feet tall in the wild. It has small and round compound leaves that alternate on the stem. The small flowers have four white petals and grow in clusters.
- Watch Out For: Make sure the water source is clean and wash picked watercress thoroughly if you plan to eat it raw. If the water is unclean or near manure, watercress can come with parasites.

To Conclude

As you can see, being familiar with botany can be a fantastic help in identifying edible plants, especially when they have lookalikes. It's also important to be able to spot the life cycle of the plant that you want, as it's better to harvest some plants at certain times. I've found many of the edible plants discussed here throughout North America, and most of them grow worldwide, so you're sure to find some close to home. Now that we've seen an overall picture of common edible wild plants that you can forage let's take a more focused look at things.

The easiest way to build up your foraging portfolio is to work season by season. With that in mind, we should begin in spring and move on from there.

6

Foraging Edible Wild Plants of Spring

Springtime is often associated with renewal, rebirth, and reinvigoration after the stagnation of winter. As Margaret Atwood said, "in the spring, at the end of the day, you should smell like dirt." I couldn't agree more. It's the perfect season for sowing and an ideal time for foraging as well. After all, what better time is there to camp out with the family and forage for wild edibles. Nature is fresh, bursting with life, and begging for us to enjoy it.

In spring, the most common wild edible plants tend to be wild greens, which are both abundant and at their youngest and most delicious. However, you can also find some edible roots and mushrooms, so keep an eye out.

Plant Profiles

Hopefully, you should be familiar with plant profiles by now. This time, we'll be looking at some of the plants that you're sure to see every spring. As usual, I'll tell you where you can find them and what to look out for.[49]

Please see the QR code for photographs of the following plants.

[49] (Waddington, 2020) (Adamant, 2018)

Stinging Nettles

Everyone is familiar with stinging nettles. Chances are, you know how to spot them because you've already brushed against one and fallen victim to their irritating sting.

This is good news. Stinging nettles aren't great to touch, but they are a tremendous edible wild plant to forage. You can find them at any time of the year, but early spring is when you'll find young, tender leaves. These are best to eat and easier to forage.

- Common Name: Stinging nettles, or just nettles.
- Scientific Name: *Urtica dioica.*
- Habitat: Can be found anywhere, as long as there is moist soil. It's especially common near bodies of water and other wet areas.
- Geographic Location: Found in temperate areas. Native to Europe, temperate Asia, and western North Africa. Nettles are abundant in northern Europe and throughout North America, except Hawaii.
- Distinguishing Features: Stinging nettles grow to a maximum of about 3 to 7 feet tall in summer, but you'll be looking for smaller, younger plants that are at most 1 foot tall in spring. You can identify stinging nettles by their leaves, which you would find on a tall, wiry green stem. The leaves are narrow and long, with a serrated edge. The stems and leaves are hairy, and these hairs usually sting.
- How to Harvest: First, always wear gloves and long sleeves, so you don't get stung. Aim for plants under a foot tall for young, tender leaves. If you can only find mature plants, only harvest the young leaves at the top. Carefully gather the top 5 leaves from each stem and keep them in a thick bag until you get back to civilization.[50] Once you get home, run them under cold water and then boil them to eliminate the sting. Once you cook them, you can use them.
- Nutritional Information: Nettles contain vitamins A, C, K, and some B vitamins. They have all of the essential amino acids and several minerals, including calcium, iron, and magnesium. Nettles also contain beta-carotene and nutrients that act as antioxidants.

---

[50] (Schatz, 2021)

- Caution: The name gives it away; stinging nettles sting. For some people, this causes an irritating rash for a few minutes, but others might have a rash for hours or even days afterward.
- Edible Part/s: The leaves are edible and are best when they're young before the plant has flowered. Nettle seeds are also edible.

Ground Elder

Ground elder is one of those invasive species that aren't native to the United States, making it the bane of many gardeners. It also makes it an excellent foraging option. However, the trick is to harvest it at the right time.

The mature leaves of the ground elder are edible, but they have a strong taste that can be unpalatable. However, if you forage the leaves when they're very young in spring, you'll find that they make a tasty, parsley-like herb.

- Common Name: Ground elder or bishop's goutweed.
- Scientific Name: *Aegopodium podagraria.*
- Habitat: Grows in shrubberies, along roadsides, on wasteland, and near rivers and streams. It likes shady, moist areas, which means that it's widespread in forests and woods.
- Geographic Location: Native to Europe and western Asia, and common in North America and Australia. In the United States, you can usually find it in northern states and most of the East Coast.[51]
- Distinguishing Features: Ground elder is a shrub that grows to between 12 and 36 inches in height.[52] You can spot it by its distinctive large, jagged leaves. The green stems are hollow inside. It has small white or reddish flowers that grow in clusters when it flowers.
- How to Harvest: Pick the leaf shoots when they're very young, ideally before the leaf has even unfurled. This will help you get the most tender and mildest leaves.

---

[51] (USDA Plants Database, n.d.)

[52] (*Ground Elder - Characteristics, Cultivation, Care and Use*, 2019), (*Ground-Elder - Aegopodium Podagraria*, n.d.)

- Edible Part/s: The young leaves are best, but you can eat the flowers, seeds, and shoots as well.
- Nutritional Information: Ground elder is particularly high in potassium and vitamin C.[53] You can use it as a medicinal herb.
- Caution: Ground elder has several toxic lookalikes, including hemlock. Make sure you know what you're harvesting.
- Lookalikes: The flowers are similar to those of the wild carrot, which means that they're also similar to water hemlock flowers. It also looks a bit like Fool's Parsley. To avoid picking the wrong plant, avoid plants that grow near water. Make sure that you select plants with green stems that grow in extensive patches. If you stick to these rules, then you'll avoid any lookalikes.

Chickweed

As the name suggests, chickweed is commonly eaten by hens. But humans like to eat it as well and for good reason. Chickweed is very common and grows almost everywhere. It's best in early spring when the leaves are nice and young and before the summer heat hits. It's one of the first things you can forage in the year.

Chickweed has a mild flavor with a nice crisp texture, making it perfect for salads or sandwiches. You can also press it against irritated skin to create a soothing poultice.

- Common Name: Chickweed.
- Scientific Name: *Stellaria media.*
- Habitat: Prefers cool and damp areas, making it common in gardens and wooded areas.[54] It grows as ground cover in meadows and fields as well.
- Geographic Location: Native to Europe and widespread throughout Asia and North America. It's more common in northern states.
- Distinguishing Features: Chickweed grows closer to the ground as coverage, with small, round leaves. If you look closely, you'll see a single line of hairs running along

[53] (Sarkar, 2012)

[54] (*Foraging in Early Spring: Wild Edible Plants to Gather Now — Good Life Revival*, 2021)

the length of each stem. It has small, white flowers with five petals with such deep clefts that they each look like they're split in half, which causes many people to count ten petals instead. The flower stem and leaves around the base of the flower are covered in fine hairs as well.

- How to Harvest: It's easier to harvest a dense patch of chickweed. Ideally, stick to harvesting the top inch or two of the stem for the best results. Use scissors, so you don't damage the plants.
- Edible Part/s: The leaves, stem, and flowers.
- Nutritional Information: Very high in vitamin C, phytosterols, flavonoids, and other beneficial vitamins and minerals.[55]
- Caution: If you eat too much chickweed, you can upset your stomach. However, you'd need to eat a considerable amount of it. Ensure that the chickweed hasn't been contaminated with herbicides or other chemicals. It does have some toxic lookalikes.
- Lookalikes: Mouse-ear chickweed isn't toxic but is completely covered in hairs. Scarlet pimpernel is a toxic lookalike. It has square stems with no hairs whatsoever. Its leaves sometimes have a reddish underside, and the flowers are red, not white.

Plantain

You might be familiar with the tropical plantain fruit, which resembles a banana. However, there are also some common weeds known as plantain. These weeds are known as the broadleaf plantain and the ribwort plantain.

I figured that I'd cover them both together because they are very similar in flavor and, as you'd expect, are closely related. You've likely seen these weeds before, so once you know how to identify them, you can harvest them at your leisure.

- Common Name: Broadleaf plantain.
- Scientific Name: *Plantago major.*

[55] (*Chickweed: Benefits, Side Effects, Precautions, and Dosage,* 2020)

- Habitat: Trampled and compacted soil; you can find it along trails and walkways. You can find it in gardens, roadsides, and disturbed areas. But plantain is very hardy and can grow in many places.[56]
- Geographic Location: Found in Europe, Asia, and North America.
- Distinguishing Features: Broadleaf plantain, unsurprisingly, has broad leaves. They grow in a rosette close to the ground. The leaves have prominent parallel veins that run the length of the leaf. You can also spot them by their central flower spike, which is covered in tiny flowers with transparent petals.
- How to Harvest: Pick the leaves from the base of the plant using either scissors or your hands.
- Edible Part/s: The entire plant is edible, but the leaves are easiest to harvest.
- Nutritional Information: High in vitamins A, C, and K, as well as zinc, potassium, and other minerals. The seeds are also good for you, rich in proteins and fatty acids.

Both of these weeds are best harvested in early spring if you want to eat them. Very young leaves are excellent in salads, but you can also use them in place of cabbage or kale. They aren't as tasty if you harvest them later in the season, but you can use them for medicinal purposes as they're very good for you.[57]

- Common Name: Ribwort plantain.
- Scientific Name: *Plantago lanceolata.*
- Habitat: Trampled and compacted soil; you can find it along trails and walkways. You can find it in gardens, roadsides, and disturbed areas. But plantain is very hardy and can grow in many places.
- Geographic Location: Ribwort plantain is particularly abundant in the Pacific Northwest.
- Distinguishing Features: Like other plantains, ribwort plantain has a rosette of green leaves with parallel veins. The leaves are narrower and shaped like pointed

[56] (*Foraging Plantain: Identification and Uses*, 2020)

[57] (*Ribwort Plantain, Narrow Leaf Plantain, Plantago Lanceolata*, n.d.), (Neverman, n.d.), (*Narrow Leaf Plantain Facts and Health Benefits*, n.d.)

spearheads. The flowers grow on thin, hairy square stems that don't stand as tall as the broadleaf plantain. You may see the tiny white flowers or the brown oval flower buds.

- How to Harvest: Pick the leaves from the base of the plant using either scissors or your hands.
- Edible Part/s: The leaves and the flower stem are edible.
- Nutritional Information: Contains a lot of vitamin C, as well as plenty of iron, calcium, and magnesium, as well as other minerals.

Miner's Lettuce

Miner's lettuce is a hardy plant, which means it starts growing even through the snow, making it a fantastic early spring forage. The leaves are thick and crunchy and not very bitter. You can use it in salads or treat it like spinach.

It has several names, including spring beauty and Indian lettuce. But the most common name, miner's lettuce, comes from the California gold rush. Miners would eat it to prevent scurvy, which shows its nutritional benefits.

- Common Name: Miner's lettuce or spring beauty.
- Scientific Name: *Claytonia perfoliata.*
- Habitat: Prefers shady and damp spots, so you can find some of the best leaves in forests under thick tree canopies. You can also find it near the coast and in fields.
- Geographic Location: Abundant in California, but common throughout the pacific northwest.
- Distinguishing Features: Miner's lettuce is made up of pairs of joined leaves, giving the appearance of a single leaf that appears spade-shaped before forming a circular round leaf. Each plant looks like a clump of 5 thin, smooth stems that each have the distinctive leaf pairing on top. A delicate flower stalk comes out from the center of the leaf pairing and forms a cluster of small, white flowers.
- How to Harvest: Be careful; miner's lettuce is fragile. Carefully pinch the stems to take the leaves alone.

- Edible Part/s: The stem, leaf, and flower.
- Nutritional Information: Particularly high in iron and vitamins A and C.

How to Use Spring Forage

It's all very well and good foraging wild edible plants in spring, but what do you do with them? You can eat some of them on the go, but that's not the best way to use most of your forage. Let's break down how you can get the most out of your favorite spring plants. I'll give you some preparation tips and even a few recipes to try out.

Using Chickweed

As mentioned before, it's best to harvest chickweed at the start of spring. Scout out likely areas for it to grow in late winter so that you can spot the first young leaves as winter gives way to spring.

You can simply add chickweed to salads, providing a fantastic nutritional boost. Chickweed can be a little bland, but it has a bright and fresh taste, which adds a springtime touch to anything you add it to.

One brilliant way to use chickweed is to make chickweed pesto.[58] You can add other wild greens to the pesto and basil to include that signature flavor. In any case, it's a perfect way to incorporate a tasty freshness into any meal. Eat the pesto with pasta, as a dressing, or as a sauce for meat or vegetables.

Chickweed Pesto Recipe

Ingredients:

- ½ cup of pine nuts, walnuts, or cashews
- 2-3 cloves of garlic, minced
- 3 cups loosely packed chickweed
- 1 tablespoon lemon juice
- ½ cup extra virgin olive oil

---

[58] (*Chickweed Pesto: Wild Greens Superfood Recipe*, 2018)

- ½ teaspoon salt
- ½ teaspoon freshly ground black pepper
- ¼ cup freshly grated parmesan cheese

Method:

1. Rinse and chop the chickweed.
2. Put the ingredients into a food processor/blender and process until smooth. You could also use a mortar and pestle, the old-fashioned way.
3. Check for taste and consistency. If it's too thick, add more olive oil. Add more lemon juice if it needs brightening.

Notes:

- Store in the refrigerator and eat within 3-4 days. You can freeze this pesto.
- This recipe also works with other foraged spring greens, like dandelions and miner's lettuce.

Using Nettles

While nettles aren't the easiest thing to harvest, they grow everywhere and are very easy to spot. This means that they are an excellent foraging option, as long as you're careful.

However, unlike other spring greens, you can't simply eat nettles as they are or add them uncooked to a salad. Well, you can try, but you certainly won't enjoy it. Nettles typically need processing before you can eat them, especially the stinging species.

Thankfully, it isn't difficult to prepare nettles for consumption.[59] Heat denatures the stinging hairs, which makes them harmless. This means that, as long as you cook nettles, you can eat them with impunity.

There are so many uses for nettles once you've prepared them. Nettles are commonly used in cheese, and you can even make nettle wine with them. A straightforward way to use nettles is to make nettle tea. All you need is nettle leaves and boiling water for a highly nutritious beverage. Add honey, lemon, or other flavorings as you like.

---

[59] (*Ultimate Guide to Wild Edibles: Spring Wild Edibles*, 2020)

However, there are also some simple dishes that showcase the wonders of nettles:

Sautéed Nettles

Ingredients:

- 1 cup of nettles
- 1 tablespoon of butter
- 1 tablespoon of fresh lemon juice, or to taste
- Salt and pepper, to taste

Method:

1. Rinse the nettles carefully, so you don't get stung.
2. Melt the butter in a pan and add the nettles. Sauté until the leaves wilt.
3. Add a squeeze of lemon juice and season to taste.

Notes:

- Serve as you would sautéed spinach or other greens. It works nicely as a side dish for chicken or pork or with a pasta dish.

Nettle & Rice Soup

Ingredients:

- 2-4 cups of nettles, washed
- ½ cup white rice
- 4 cloves of garlic, minced
- ½ tablespoon of turmeric powder
- 1 teaspoon of salt
- 1 tablespoon of butter
- 4 cups of water or stock
- Freshly ground black pepper (optional)
- Squeeze of fresh lemon juice (optional)

Method:

1. Boil the nettles, rice, turmeric, and salt in the water until the rice is cooked and the nettles are nice and tender. It should take about 15-20 minutes. Add more liquid if needed.
2. Heat the butter and gently sauté the garlic.
3. Add the garlic butter to the soup and taste. Adjust the seasonings if necessary. Add the black pepper and lemon juice to suit your taste. Serve in bowls and enjoy.

Notes:

- This soup makes a great lunch or starter course. It's healthy and, because of the rice, you don't even need to bother with bread.

To Conclude

Spring is the perfect time to harvest fresh greens. You can easily identify most of these greens and simply forage them as you go. They aren't just tasty, but they're incredibly healthy for you. You just need to know how to use them. Now, let's have a look at the great stuff that you can find in summer.

7

Foraging Edible Wild Plants of Summer

Summer is a fantastic season, and spending your summer outside can have some great health benefits.[60] The warm sun brings vitamin D and encourages activity. But, best of all, summer brings wonderfully varied opportunities for foraging. So, while you're outside enjoying the fine summer weather, you might as well take advantage of the wild berries, fruits, greens, and herbs that are waiting for you.

Plant Profiles

There's such a variety of things to find in summer, so we'll break down the plant profiles logically. Whether you're looking for fruit, flowers, roots, or seeds, you'll know exactly what to look out for.[61]

Please see the QR code for photographs of the following plants.

Fruits and Berries

You'll find some great fruit and berries in late summer and early fall. Here are a couple of my favorites.

---

[60] (*6 Surprising Ways Summer Can Be Good For Your Health*, 2020)

[61] (*The Beginner's Guide to Late Summer Foraging*, 2011)

Raspberries

There are two species of wild raspberries that you're likely to find in North America.[62] One, the American red raspberry, looks like the kind of raspberry that you'd find in the wild. But you can also find black raspberries, which, as you'd expect, are black when they're ripe.

- Common Name: American raspberry and black raspberry.
- Scientific Name: *Rubus strigosus and Rubus occidentalis.*
- Habitat: Raspberries like rich soil, as you'd find along field edges and old farmland. You can also find them in disturbed areas, meadows, and near bodies of water.
- Geographic Location: You can find these two species throughout North America. They prefer colder climates, so you're more likely to find them further north. You can find other species in other parts of the world, and they all share some distinctive features that make them easy to spot.
- Distinguishing Features: Raspberries grow in canes up to about 5-6 feet long, sometimes upright and sometimes arched. The large leaves grow in groups of 3 or 5. The leaf edges are toothed, and they come to a sharp point. Sometimes the leaves have a white underside. Raspberry flowers appear in spring, with five white petals. In early summer, immature compound berries will develop, either green or white in color. Depending on the species, they will ripen to a pinkish-red or dark purple color. The berries detach easily from the plant, and you might notice tiny fine hairs.
- How to Harvest: Gently pull the berries. They should detach with very little resistance, but be careful not to crush them. Ripe berries can be delicate.
- Edible Part/s: The berries and leaves are edible.
- Nutritional Information: They're a great source of fiber, vitamin C, and manganese, as well as other nutrients.[63]
- Caution: The leaves shouldn't be consumed in early pregnancy. Raspberries do have thorns, so be careful.

---

[62] (Suwak, 2018) (Linnaeus, n.d.) (*Rubus Strigosus*, n.d.)

[63] (*Red Raspberries: Nutrition Facts, Benefits and More*, 2018), (*Foraging Guide Raspberry | UK Foraging*, n.d.)

- Lookalikes: Blackberries look a bit like black raspberries. Red blackberries look like thimbleberries. Both of these berries are edible.

Wild Plums

There are about 30 different varieties of wild plums in the United States[64], and thousands of different varieties around the world. I'll mainly be discussing the most common wild plum in North America, but you can use the same tips for other varieties.

- Common Name: Wild plum or American plum.
- Scientific Name: *Prunus americana.*
- Habitat: Commonly found along fencerows and roadsides or in open fields. They can thrive in many different environments, so don't be surprised if you find plums in savannas, swamps, on rocky hillsides, or near wooded areas.
- Geographic Location: You can find this species in the American Midwest. Other types of plum grow in Canada, the West Coast, or further south. Plums are common throughout the Northern Hemisphere.
- Distinguishing Features: Plum trees are relatively small, growing as shrubs or trees about 15 feet tall. The branches are thin and covered in a rough, reddish-brown bark. Plum trees may have distinctive large thorns up to 3 inches long. The pointed leaves are finely serrated and have a wrinkled, dark green upper side that contrasts with a smooth, pale green underside. Plum flowers appear in mid-spring in small clusters. They're small, with five white petals that have a strong sweet smell. Some plums are ripe as early as May, while others ripen in late summer and early fall. The small, oval fruit can range from reddish-yellow to purple, depending on the variety. They may be covered in white powder when ripe.
- How to Harvest: You have to pick plums when they're ripe because they don't last long. They should easily come away from the tree. Scan the ground for any good plums that haven't yet decayed.
- Edible Part/s: The fruit of the plum tree is edible.

---

[64] (Weekes, n.d.), (*Wild Plum Facts and Health Benefits*, n.d.)

- Nutritional Information: Plums are very high in vitamins A, C, and K. They have a bit of vitamin B5 and B6 in them as well. They also contain a lot of potassium.
- Caution: Watch out for the thorns. They're hard to miss but no joke if you fall into them.
- Lookalikes: Different species of plum trees are often confused with one another, but they're all edible. Some do differ in taste.

Flowers

Summer blooms are stunning, but they're more than just a pretty face. Edible flowers are great to add to salads, use as garnishes, eat as snacks, or turn into tea blends.

Yarrow

Yarrow is a fantastic foraging option, as it's very common and healthy. It's often used for medicinal purposes.[65] For example, yarrow tea has anti-inflammatory properties and is commonly used to lower a fever. It tastes bitter, but it's so valuable and versatile that it's more than worth it.

- Common Name: Yarrow.
- Scientific Name: *Achillea millefolium.*
- Habitat: You can find yarrow in most environments, most commonly in meadows, fields, and disturbed areas.
- Geographic Location: Grows in temperate climates all around the world.
- Distinguishing Features: Yarrow has distinctive frilly, feather-like leaves. The grooved, slightly hairy stems grow to about 2-3 feet high and are topped with a small cluster of flowers, which are usually white or pink. Some plants can have other colored flowers. Yarrow smells like fresh pine needles when crushed.
- How to Harvest: Yarrow can be harvested in spring and summer. Simply snip the uppermost stems to get the young leaves and flowers. You can also harvest the leaves.

[65] (*5 Emerging Benefits and Uses of Yarrow Tea*, 2019), (*Foraging Yarrow: Identification, Lookalikes, and Uses*, 2021), (Larum, 2021), (Rey, n.d.)

- Edible Part/s: The whole plant is edible. The leaves and flowers are most commonly used.
- Nutritional Information: Yarrow contains vitamins A and C, as well as potassium, zinc, magnesium, and other minerals.
- Caution: Pregnant women and nursing mothers shouldn't consume yarrow. It's sometimes used to relieve menstruation, but this means that it can interfere with pregnancy.
- Lookalikes: The clusters of small flowers can cause yarrow to be mistaken for wild carrot (Queen Anne's Lace), and our old friend poison hemlock. Wild carrot is edible; poison hemlock is assuredly not. Poison hemlock is much larger than yarrow, and the leaves are different. If nothing else, check the leaves for that distinctive pine scent.

Chicory

Chicory is one of those handy plants that are entirely edible. The leaves and flowers can be eaten raw in salads, although some prefer to cook the bitter leaves. The roots are beneficial, as you can roast them, ground them, and use them as a caffeine-free coffee substitute.[66]

- Common Name: Chicory.
- Scientific Name: *Cichorium intybus.*
- Habitat: Prefers roadsides and open fields.
- Geographic Location: Native to the Mediterranean, but spread worldwide.
- Distinguishing Features: The most distinctive feature of the chicory plant is its sky blue or sometimes light purple flowers. The petals are long and thin, with a blunt, toothed end. The plant can grow to almost 5 feet tall, with a tough, hairy stem that stands upright. The leaves are narrow and irregularly toothed. Chicory has a brown taproot.
- How to Harvest: To get at the root, dig around the plant with a tool to loosen the soil until you can dig out part of the root.
- Edible Part/s: Flowers, roots, and leaves.

---

[66] (*Foraging for Chicory*, 2016), (House, n.d.)

- Nutritional Information: Chicory root is an excellent source of fiber and prebiotics. The leaves are also very healthy, with significant amounts of vitamins K, A, C, and some B vitamins. They also contain manganese and moderate amounts of Vitamin E and calcium.[67]
- Caution: Vehicle pollution can contaminate chicory near roadsides, so make sure they're some distance from the road itself.

Roots and Bulbs

If you have your trusty trowel or another digging tool, you can take advantage of some great summer roots and bulbs. Just make sure the rules in the area allow you to dig up plants first.

Horseradish

If you like mustard or wasabi, then you'll love horseradish. This fiery plant is likely native to western Asia but has found popularity in Europe and North America. Horseradish is often cultivated, but it can run rampant and grow wild.[68]

- Common Name: Horseradish
- Scientific Name: *Armoracia rusticana.*
- Habitat: Wasteland, meadows, and other wild grassy areas.
- Geographic Location: Horseradish is native to West Asia and the Mediterranean and spread throughout Europe. It can also be found wild throughout North America, mainly where it's cultivated in Illinois, Wisconsin, and California.
- Distinguishing Features: Horseradish has long stalks with white, sweet-smelling flowers on a spike. The leaves are large, slightly toothed, and resemble dock leaves in shape. They have a shiny green color. The knobbly taproot is brown and has a pungent odor.
- How to Harvest: Use a sharp knife or scissors to forage the leaves and flowers. Dig up the root to harvest it. Horseradish is hardy, but don't be too destructive.

---

[67] (*Chicory*, n.d.)

[68] (Linnaeus, n.d.), (*Horseradish - A Foraging Guide to Its Food, Medicine and Other Uses*, n.d.),

- Edible Part/s: The root is most commonly eaten, but the leaves and flowers are edible too.
- Nutritional Information: High in vitamin C and has a decent amount of fiber, which isn't easy to find in foraged food.
- Caution: Too much horseradish can irritate the gut. Be cautious if you're pregnant, as some sources suggest that horseradish can endanger the fetus.

Wild Garlic

Wild garlic, otherwise known as field garlic or crow garlic, is native to Europe and the Middle East. It's considered a noxious weed in North America. Confusingly, there are other species of edible alliums known as wild garlic, but you can harvest this species for its bulbs in summer.[69]

- Common Name: Wild garlic, field garlic, or crow garlic.
- Scientific Name: *Allium vineale.*
- Habitat: As the name suggests, field garlic is commonly found in fields. You can also find it in meadows, disturbed areas, grassy wastelands, and woodland edges.
- Geographic Location: In North America, it's common in the midwestern states. You can also find it in Europe, Northern Africa, the Middle East, and Australia.
- Distinguishing Features: This plant grows in bunches of long, narrow, pointed leaves that are bluish-green in color and hollow. They can grow between 1 foot to 3 feet tall. The leaves smell strongly of onions.
- How to Harvest: You can sometimes gently pull the whole plant close to the ground, which will let you harvest the leaves and bulb alike. Otherwise, use a digging tool or cut them if you plan on only harvesting the leaves.
- Edible Part/s: The whole plant is edible. The leaves can be used as herbs, while you can use the roots similarly to garlic.

---

[69] (Linnaeus, n.d.), (*How to Identify Field Garlic - Foraging for Wild Edible Plants — Good Life Revival*, 2017), (*Field Garlic Information and Facts*, n.d.)

- Nutritional Information: Contains vitamins A and C, as well as potassium, manganese, calcium, and selenium. It also has immune-boosting properties, like other alliums.
- Caution: Be careful of foraging in active fields, as wild garlic may have been treated with herbicides.
- Lookalikes: The leaves look like chives, which are also edible.

Nuts and Seeds

Nuts and seeds are more commonly found in fall, but you'll find some delicious treats in summer as well.

Carrot seeds

Queen Anne's lace or wild carrot is one of those plants that can be foraged all year round.[70] You can harvest the roots in spring when it's tender, and you can harvest the whole plant, or just the seeds, in late summer and early fall.

- Common Name: Wild carrot or Queen Anne's lace.
- Scientific Name: *Daucus carota.*
- Habitat: Usually found near the coast on cliffs and dunes, but can also be found in meadows and rough grassland.
- Geographic Location: Native to temperate areas in Europe and southwest Asia and common throughout North America and Australia. You can usually find it in the eastern states of the US, as well as the south and west coasts.
- Distinguishing Features: The plant is about 1 to 3 feet tall and has a stiff, hairy stem. The leaves are lacy and triangular. The white flowers bloom in clusters, and the center few flowers are sometimes dark or red. The flowers sit above 3-forked leaves. When they go to seed, the cluster closes in on itself and dries, making a tumbleweed-like structure.

---

[70] (Linnaeus, n.d.), (*Queen Anne's Lace - The Wild Carrot*, n.d.), (Haughton, n.d.)

- How to Harvest: The root should be dug up in spring when it's less tough and fibrous. You can harvest the seed when the flowers dry up by cutting off the flower head and up-ending it over a paper bag. Shake the flower head, and it will release the seeds.[71]
- Edible Part/s: The seeds, flowers, and roots are edible, although the root is unpalatable.
- Nutritional Information: Wild carrot seeds have traditionally been used for medicinal purposes, including for digestive problems and even as a contraceptive.
- Caution: Has toxic lookalikes. Sometimes, the leaves cause some people to have a mild reaction when touched.
- Lookalikes: Looks similar to poison hemlock. Hemlock plants are usually larger and don't have forked leaves beneath the flowers or dark flowers. Hemlock also has an unpleasant smell, while wild carrot smells like carrot.

Seaweed

If you live near some bodies of water, then keep an eye out for some tasty seaweed.

Dulse

Dulse is commonly known as a superfood.[72] Like most seaweed, dulse is salty, but it has a relatively mild flavor. You can harvest it all year round, but because it grows in colder, northern climates, most foragers prefer to look for it in warm summer.

- Common Name: Dulse.
- Scientific Name: *Palmaria palmata.*
- Habitat: Prefers cold saltwater in areas with strong currents.
- Geographic Location: Grows on the eastern shore of North America, from Canada to New Jersey.

---

[71] (Palomo, n.d.)

[72] (Demers, 2021)

- Distinguishing Features: Dulse is dark brown and purple. It can grow to between 6 inches to 18 inches in length. It has divided lobes and flat, irregularly branched fronds. It can sometimes latch itself to other underwater plants.
- How to Harvest: Wait until the low tide so you don't get caught in the current. Then, look along the exposed rocks for the seaweed.
- Edible Part/s: The fronds.
- Nutritional Information: Very rich in iodine, vitamin B6, iron, potassium, fiber, and protein. It also has anti-inflammatory properties.
- Caution: Dulse is very popular, so don't take too much. Also, make sure that there are no nearby chemical runoffs. Dulse is rich in iodine, so eat it in moderation. When harvesting, don't get caught out when the tide comes back in.
- Lookalikes: Another seaweed called laver looks similar to dulse, but it's also edible.

How to Use Summer Forage

Some summer forage, such as berries and fruit, are delicious when you snack on them as you go. But other edible plants are far better when prepared and cooked as ingredients. Here are some ways to make the most of them.

Using Yarrow

As mentioned earlier, yarrow has some fantastic medicinal benefits and is often used as a tea.[73] While you can simply throw some leaves and flower heads in with boiling water, you may find the result unpalatably bitter. You can also make a tincture, which provides the same health benefits, and you can take it in small doses.

Yarrow Tincture

Ingredients:

- Handful of yarrow flowers
- Cup of vodka

---

[73] (Long, 2019)

Method:

1. Put the yarrow flowers in a glass container and cover with alcohol.
2. Leave for six weeks.

Notes:

- A typical dosage for this tincture is 2ml twice a day. Add it to a hot beverage.

Yarrow Tea

Ingredients:

- 1-2 teaspoons of dried yarrow flowers and leaves
- 1 cup of boiling water
- Honey and a lemon slice, for taste

Method:

1. Put the dried yarrow in a teapot with the boiling water. Brew for up to 20 minutes.
2. Strain into a cup and add honey and the lemon slice for taste.

Notes:

- If you have loose-leaf tea, you can add dried yarrow to the mixture.

Using Horseradish

The most common way to use horseradish[74] is to make a sauce. You can also pickle it. Generally, it's best to store horseradish root whole in the refrigerator and grate it whenever you want to use it.

Horseradish Sour Cream

Ingredients:

- 2-3 tablespoons of grated fresh horseradish
- 1 cup of sour cream

---

[74] (Albert, n.d.)

- 1 tablespoon of lemon juice
- Salt, to taste

Method:

1. Mix 2 tablespoons of horseradish with the sour cream.
2. Taste, then add the other ingredients as you'd prefer.

Notes:

- Traditionally served with beef or asparagus.

To Conclude

Summer is a fantastic time to forage a tremendous variety of fresh plants. I've only touched on a few options here, so it pays to do additional research about your specific area. If you spot a promising plant, then take note of it. Next, let's look at some of the wonders of fall.

8

Foraging Edible Wild Plants of Fall

One of my favorite quotes about fall, penned by the classical novelist Samuel Butler perfectly explains why fall is a forager's dream. "Autumn is the mellower season, and what we lose in flowers we more than gain in fruits." In fall, there's something edible practically wherever you look. But while you're enjoying the abundance of fall, keep in mind that you should only take what you need and forage responsibly. Good foraging groves are gold if you can keep them going for another year.

Plant Profiles

Most flowers have gone to seed by fall, which means plants produce fruit and nuts in abundance. Here are some that I like to forage whenever I spot them.[75]

Please see the QR code for photographs of the following plants.

Apples

Wild apples grow in abundance in fall. You can get different species throughout North America and the rest of the world. Small, sour crabapples[76] are most common, but you'll also find larger, sweeter varieties in the wild. Different apples are better suited for cooking, while

---

[75] (*What to Forage in Fall: 30+ Edible and Medicinal Plants and Mushrooms*, 2018), (Mullins, 2014)

[76] (Deane, n.d.)

others can be eaten raw. Each apple tree can be slightly different, so you'll have to test each one before knowing how best to use it.

- Common Name: Apple or Crabapple.
- Scientific Name: *Malus*.[77]
- Habitat: Apples can commonly be found in thickets, near roadsides, and on rough, abandoned ground. You can also find them in copses or in wooded areas near hills.[78]
- Geographic Location: Apples grow worldwide, especially in the Northern Hemisphere. Crabapples are native to North America and Europe, while domesticated sweet apples originated from central Asia.
- Distinguishing Features: Apple trees can range between 13-40 feet tall and tend to have a dense, twiggy crown. The long, simple leaves grow alternately and usually have serrated edges. The flowers have five petals and can range from white, pink, or red in color. The stamens are generally red. They can grow as small, oval-shaped crabapples or larger, round apples when they fruit.
- How to Harvest: When plucking from the tree, twist the apple gently until it comes away from the branch or use a sharp tool. Don't just yank on the tree as you risk breaking the branch. You can find good apples on the ground, but you should cook them rather than eat them raw.
- Edible Part/s: The apples themselves are edible.
- Nutritional Information: Apples are high in fiber and have a wide variety of vitamins and minerals. They have decent amounts of vitamin C and potassium.[79]
- Caution: Apples are popular with bugs, particularly wasps and maggots. Check each apple for bruises or damage and especially holes where insects may have eaten their way in.

---

[77] (Miller, n.d.)

[78] (*Apple (Malus X Domestica) - British Trees*, n.d.)

[79] (Arnarson, 2019)

Persimmons

Persimmons are delicious and are best when they're perfectly ripe and soft. If you eat them at the wrong time, you'll find that they're sour or bitter. But a ripe persimmon is incredibly sweet, even more so than commercially available Asian persimmons.

- Common Name: American persimmon.[80]
- Scientific Name: *Diospyros virginiana.*
- Habitat: Prefers sandy soil in full or partial sun. It may be one of the few trees that grow in that area, as this habitat is less hospitable to other plants.
- Geographic Location: Grows in the southeastern states of the USA, ranging from Connecticut to Florida and as far west as Texas.
- Distinguishing Features: The trees are tall, reaching up to 35 to 60 feet. They have distinctive rough, gray-black bark that looks almost like crocodile skin. The teardrop-shaped leaves are glossy and between 4 to 8 inches long. The fruit looks a bit like a small, bright orange plum about 2 inches in diameter and has prominent leaves at the stem end.
- How to Harvest: Harvest the fruit when it's slightly wrinkly and has either fallen from the tree or detaches easily. It should be very soft when you eat it, but you can harvest unripe persimmons as long as they're orange and they can ripen at home.
- Edible Part/s: The orange fruit is edible.
- Nutritional Information: Persimmons are very nutritious, with a considerable amount of vitamins A and C and plenty of manganese. They also have other vitamins and minerals to offer.
- Caution: Persimmons must be perfectly ripe (bordering on overripe) before eating, or you'll find them unpalatable.

[80] (Linnaeus, n.d.), (*Top 7 Health and Nutrition Benefits of Persimmon*, 2018), (Meredith, 2014)

Rosehips

The rose flower, especially the wild rose, produces plenty of edible fruits known as rosehips. They're rarely eaten raw, but you can use them in plenty of delicious recipes. The best thing is that they're easy to spot and abundant on the plant.

- Common Name: Rosebush[81]. Every rose produces rosehips.
- Scientific Name: *Rosa.* The wild rose is *Rosa acicularis.*
- Habitat: Wild roses are very hardy and can grow in many different environments, but you'll be more likely to find them in wet soil. You usually find them in thickets, near streams, and on wooded hillsides.
- Geographic Location: Most species are native to Asia, but roses are common worldwide.
- Distinguishing Features: Grows in a 3 to 9-foot tall shrub. The ovate leaves have toothed edges and grow in groups of 3 to 7. The flowers of wild roses are usually pink and create red, elongated rosehips less than an inch in diameter.
- How to Harvest: Harvest in late fall and early winter. Some say they're nicest after the first frost, but foraging early means beating other critters. Pull or cut the rosehips directly off the canes. They should be red or orange and firm.
- Edible Part/s: You can use the rosehips themselves in recipes. Some people make tea from the leaves.
- Nutritional Information: Rosehips are very high in vitamin C and A, to the point that in Britain, they would use the syrup as a vitamin supplement.
- Caution: Roses usually have sharp thorns, and wild roses can be particularly vicious. Rosehips also contain hairy seeds that can irritate the skin (they used to be used by mischievous children to make itching powder), so be sure to remove the hairs when you process them.

[81] (*Rose Hips: When, How, and Why to Harvest*, 2016), (*How to Harvest and Use Rose Hips - Flowers*, 2021), (*Rosa Acicularis*, n.d.)

Blackberries

Blackberries are inexorably linked with fall, most people can identify them at a glance, and you've likely foraged them before. They're delicious raw or prepared with other fall forage and have no toxic lookalikes, making them an excellent option for children and inexperienced foragers.

- Common Name: Blackberry bush.
- Scientific Name: *Rubus.* There are numerous species throughout the world.
- Habitat: Common in woods, near roads, or wasteground.
- Geographic Location: Native to northern temperate areas, particularly in eastern North America.[82]
- Distinguishing Features: Depending on the species, blackberry bushes may form erect canes while others trail across the ground. Most blackberry bushes have thorns or prickles along the canes or stems. The rounded, toothed leaves grow in groups of 3 or 5. The flowers are white, pink, or red and produce distinctive black compound berries. Unlike raspberries, which are hollow, blackberries have a firm white core.
- How to Harvest: Gently twist each blackberry; it should come away with little resistance. Avoid any that are shriveled. If the berries are challenging to pick, they likely aren't completely ripe, even if they're black or purple.
- Edible Part/s: The berries are edible, and you can make the leaves into tea.
- Nutritional Information: Full of antioxidants, as well as vitamins C, E, and K. They also contain iron, calcium, and manganese.
- Caution: Those thorns can cause deep scratches. Be careful and wear long sleeves to avoid any injuries. Resist the urge to go for blackberries out of your reach, as you may fall into the bush.
- Lookalikes: Has several edible lookalikes, including dewberries that grow along the ground and black raspberries.

---

[82] (*Blackberry | Fruit | Britannica*, n.d.), (Brennan, 2020)

Chestnuts

Unfortunately, American chestnut trees are now critically endangered because of chestnut blight, which was introduced in 1904. However, you can still find some if you know what you're looking for, and the sweet, delicious chestnuts are more than worth it.[83] Other species also grow in North America and Eurasia.

- Common Name: American Chestnut or sweet chestnut.
- Scientific Name: *Castanea dentata.*
- Habitat: You can find chestnuts in wooded areas and small chestnut groves. You can sometimes find a single mature chestnut tree in a meadow or field where the blight hasn't exposed it.
- Geographic Location: Northeastern United States, particularly Wisconsin.
- Distinguishing Features: It can be tricky to find mature trees, but it's worth keeping an eye out for them. They are tall, many reaching over 50 feet in height. The leaves are long and skinny, with pointed teeth along the edge. You can find the chestnuts themselves in distinctive spiny green husks or burrs that grow in clusters.
- How to Harvest: Harvest between mid-September to November and look for the spiny burrs on the floor. The burrs should be slightly open. Discard any with any signs of damage or holes.
- Edible Part/s: The chestnuts themselves are edible. You can eat them raw or cooked.
- Nutritional Information: Chestnuts are very rich in vitamin C and are a good source of antioxidants.[84]
- Caution: When foraging, take special care not to damage the tree or overharvest. Forage nuts on the ground and leave any with holes. Wear gloves, as chestnuts have spiny casings.

---

[83] (Demers, 2021) (*American Chestnut*, n.d.) (Huffstetler, 2021)

[84] (Brennan, 2020)

- Lookalikes: The horse chestnut is toxic and can be mistaken for sweet chestnuts. It's easiest to check the spines. Horse chestnuts are larger and have fewer, short spines, while sweet chestnuts have many fine spines.

Beechnuts

Beechnuts are an underrated wonder. They aren't available every year, as the trees only produce full nuts every 3-5 years. But, when they're ready, and you know where to find them, they're abundant. You will need to shell them before eating, as with other nuts.

- Common Name: American beech.[85]
- Scientific Name: *Fagus grandifolia.*
- Habitat: Beech trees prefer areas where they have access to water. You can find them in wooded areas and fields.
- Geographic Location: The American beech is typically found in eastern North America, but the European beech also grows in North America. Other beech trees grow in Europe and Asia.
- Distinguishing Features: The beech tree has smooth, silver-gray bark. It can grow to about 50 to 70 feet high. It has a low, dense canopy of toothed, dark green leaves. You will likely spot the brown, spined beech burrs on the ground in fall.
- How to Harvest: You'll find the spiky husks on the ground after late September. Each husk contains two small nuts. Some beechnut shells are empty, so avoid any that are slightly concave or collapsed on one side.
- Edible Part/s: The small, cream-colored nuts are edible and safest when cooked.
- Nutritional Information: Beechnuts have a great deal of protein and fat, which can round out a forager's diet. They also contain copper, manganese, and several B vitamins.[86]
- Caution: You shouldn't eat the nuts when unripe. Even ripe nuts should be consumed in small amounts when raw but are fine when roasted. Beechnuts grow in spiny

[85] (Adamant, 2018), (Myers, 2022), (*American Beech*, n.d.),

[86] (*Beechnut - Dried Nutrition Facts | Calories in Beechnut - Dried*, n.d.)

husks, which aren't sharp but can be uncomfortable. Remove any husks or skins before eating the creamy-colored nuts.

How to Use Fall Forage

While you can eat plenty of fall forage as you go, you can make some wonderful things with a little bit of effort. Here are some recipes that I like to play around with on those cool fall evenings.

Using Blackberries

Blackberries are delicious raw, but they only last for a couple of days after being picked. If you aren't going to eat your haul within that time or freeze it, you should use it before it spoils. A traditional way to use blackberries is to bake them into pies along with other fall fruit, such as apples. You can also use it to make red wine. But here are some lesser-known blackberry recipes[87] that are no less delicious.

Pickled Blackberries

Ingredients:

- 2 lb of blackberries
- 1 lb of granulated sugar
- 8½ fluid ounces of white wine vinegar

Method:

1. Rinse the blackberries and wash away any debris.
2. Boil the sugar in with the vinegar.
3. Once the sugar has dissolved, add the blackberries.
4. Simmer until the blackberries are soft, remove them, and put them into sterilized jars.
5. Continue to cook the syrup until it's thick, then pour it over the fruit.

---

[87] (*A Beginner's Guide to Autumn Foraging | Live Better*, 2014), (*Blackberry Leaf Tea: A Herbal Remedy For Your Health*, 2021)

6. Seal the jars and store for at least a week before cracking them open.

Notes:

- You can add a rose geranium leaf or dried rose petals to the jar before sealing for a subtle aromatic taste.
- These blackberries are delicious with cheese and meat, and the vinegar syrup works well in dressings or even on ice cream.

Blackberry Leaf Tea

Ingredients:

- 1 heaped teaspoon of dried blackberry leaves
- 1 cup of water
- Honey, for taste

Method:

1. Boil the water.
2. Put the dried leaves into an infuser or bag and then into a cup.
3. Pour the boiling water into the cup and let it steep for about 5 minutes.
4. Remove the tea and add honey to sweeten.

Notes:

- You can also add other dried leaves or tea to make a blend. Or, throw some fresh blackberries into the mix.

Using Rosehips

You can try to eat rosehips off the plants, but you won't enjoy it very much. Instead, it would be best if you processed rosehips before eating them. I'll tell you how to get rid of the irritating hairy seeds, as well as a classic and delicious recipe for rosehip syrup.[88] You'll get to see what those British WW2 households were raving about.

Processing Rosehips

Ingredients:

- Rosehips

Method:

1. Wash the rosehips, then use a sharp knife and a great deal of care to cut off the blossom and stem ends.
2. Cut the rosehips in half, then use a butter knife to scrape out the hairs and seeds. Discard them, along with any mushy or spoiled ones, during this step.
3. Give them a final rinse to wash away any remaining hairs.

Notes:

- You can then freeze, dry, or use the rosehips at your leisure.

Rosehip Syrup

Ingredients:

- 1 cup of rosehips, preferably prepared
- 2 cups of water
- 1 cup of honey

Method:

1. If you haven't prepared the rosehips, put them into a food processor and break them up.
2. Put the rosehips and water into a saucepan and bring them to a boil. Lower the heat to medium-low and simmer for 15 minutes or until the water has reduced by half.

---

[88] (*Radiant Rose Hips: How to Harvest, Dry and Use Rosehips*, 2022), (*Rose Hip Syrup: Foraged and Made With Honey*, 2020)

3. Take the pan off the heat and let the mixture cool down. Mash it with a potato masher to extract more juice.
4. Strain out the rosehips through a fine-mesh sieve, ideally lined with cheesecloth. Squeeze the mash to get the most out of it. If you've already prepared the rosehips, this shall do. If not, repeat this step a few more times to get rid of any hairs.
5. Once the liquid has cooled to room temperature, add the honey.

Notes:

- Store the syrup in a refrigerator. It'll keep for up to 6 months or longer if frozen.

To Conclude

Fall is the perfect time to find plenty of nuts and berries, as well as some mushrooms (we'll touch on those later). It's a little colder but no less stunning and enjoyable. Next, let's look at some of the things you can find in winter.

9

Foraging in Winter

When many people think of winter, they consider it a season of hibernation or a time to wait for spring's new life and warmth. It doesn't seem like the ideal time to be a forager, but you could view it as a fantastic opportunity to develop your foraging skills and knowledge. John F. Kennedy expressed the virtues of doing things "not because they are easy, but because they are hard."

Winter isn't just a challenge, though. You can find fantastic foraging opportunities, even beneath the snow and among the long-dormant plant life.

Plant Profiles:

Here are some of the tastiest things that you can find even in the depths of winter. While winter foraging can get chilly and uncomfortable, it's more than worth the effort.[89]

Please see the QR code for photographs of the following plants.

Cranberries

Cranberries are one of the quintessential winter berries, as the bright red berries stand out against the white snow. However, you'll likely spot two cranberry-like berries on your foraging travels.

---

[89] (Adamant, 2018), (*What to Forage in Winter: 30+ Edible and Medicinal Plants and Fungi*, 2020)

True cranberries are tart and slightly sweet. While they usually first fruit before winter, the snow keeps them perfectly preserved until the spring thaw. This means that if you know where the cranberries are beneath the snow, you can harvest them throughout the winter.

- Common Name: Cranberries.
- Scientific Name: *Vaccinium macrocarpon.*
- Habitat: Prefer wet, acidic soil. Can be found in bogs, swampy areas, coastal areas, and pine barrens.
- Geographic Location: You can find this species in eastern Canada, Northeastern states of the USA, upper Midwestern states, and south of North Carolina. You can find other species in colder parts of the world.
- Distinguishing Features: True cranberries trail along the ground, creating a dense mat of woody, slender stems. The stems rarely branch and grow to only a foot tall. The evergreen leaves are leathery, forming ½ inch ovals with blunt tips and pale undersides. In the spring, small, pale-pink flowers with four petals grow. In fall, the berries ripen to a rich red color. They look large against the stems, but they're light and seem hollow.
- How to Harvest: Identify cranberry plants in summer and fall so that you know where they are in winter. Then, simply reach underneath the frost and uncover the frozen berries. Gently twist them from the stems and let them thaw at home.
- Edible Part/s: The berries are edible.
- Nutritional Information: Cranberries are very high in vitamin C, as well as beta carotene and anthocyanins. They are also high in pectin, making them suitable for jellies, jams, and thick sauces.
- Caution: They're very close to the ground, which means some animals may have urinated near them. Rinse your cranberries.
- Lookalikes: Dogwood and holly bushes also produce bright red berries in late fall and throughout the winter, but these berries are poisonous.[90] Thankfully, both plants have

[90] (MacWelch, 2019)

much larger leaves than cranberries or highbush cranberries. The toxic berries are usually smaller as well.

Highbush cranberries

Aren't actually cranberries, but they look and taste very similar. There are three main species that you'll encounter. The European highbush cranberry (*Viburnum opulus*) is very bitter, while American highbush cranberries (*V. trilobum*) and squashberries (*V. edule*) are deliciously bright and tart. Most highbush cranberries are generally best when cooked.

- Common Name: Highbush cranberries.
- Scientific Name: *Viburnum (opulus/trilobum/edule)*
- Habitat: These plants don't need wet soil but are commonly found along streams and ponds. You can often find the European highbush cranberry near towns and farmland. You can find the American highbush cranberry near water and in wooded areas and squashberries in boreal forests.
- Geographic Location: The European highbush cranberry is native to Europe but is commonly found in southeastern Canada and eastern states. American highbush cranberries are native to the northern states and the southern third of Canada. You can find squashberries in far northern states, Canada, and Alaska.
- Distinguishing Features: The three species look similar, and the easiest way to tell one from another is to taste it. They grow in bushes about 15 feet high, with large, three-pronged maple-shaped leaves. They have smooth, white-colored bark. The small, round berries are bright red. The berries contain a single large seed, which is bitter when chewed on, unlike cranberries.
- How to Harvest: Simply pick the berries.
- Edible Part/s: The berries. The bark can be used to soothe menstrual cramps, but don't damage the bush if you can help it.
- Nutritional Information: Very high in antioxidants, vitamin A, vitamin C, and fiber.
- Caution: In large quantities, raw highbush cranberries can cause stomach issues. They usually taste better cooked anyway, especially if you find the bitter European ones.

- Lookalikes: Dogwood and holly bushes look more similar to highbush cranberries. However, you can still use the leaves to tell the difference. If the leaves aren't shaped like broad maple leaves, leave them alone.

Juniper Berries

Juniper berries[91] are most famously associated with gin, but they're used to flavor food as well. They're especially delicious with rich, fatty, or gamey meats. Despite the name, juniper berries aren't berries at all but are modified cones. If that wasn't confusing enough, some species that are commonly called cedars (like the eastern red cedar and the southern cedar) are actually junipers. Strange, I know. I'll be focusing on the common juniper, as it's very widespread, and some other juniper species are toxic.

- Common Name: Juniper bush or tree.
- Scientific Name: *Juniperus communis.*
- Habitat: Found in chalky soil and rocky ground. Junipers like to grow in moorlands, lowlands, and pine woodlands. It can thrive in high-altitude areas.
- Geographic Location: This species thrives in cool temperate areas around the world. In North America, you can find it in Canada and the northern states.
- Distinguishing Features: Grows as either a spreading shrub or a small tree. It has needle-like green leaves that grow in whorls of three. The leaves have a single white band on the inner surface. You can identify junipers by the berry-like cones. They are initially green, but after 18 months, they ripen to a purple-black color with a waxy coating. Each berry/cone contains 3 or 6 seeds within fused scales. The trees have an aromatic scent that may remind you of gin.
- How to Harvest: Avoiding the spines, carefully pull the berries from the plant. Wear long sleeves and gloves. You can also harvest a shrub by shaking it over a tarp, then grabbing the ripe berries.[92]

---

[91] (*Foraging Juniper Berries for Food and Medicine,* 2020), (*Junipers - Eat The Weeds and Other Things, Too,* n.d.), (Linnaeus, n.d.)

[92] (Grant, 2021)

- Edible Part/s: The berries. The sap, spines, and bark of some junipers are edible, but your best bet is the berry itself.
- Nutritional Information: Rich in vitamin C. Juniper berries are known to contain an antiviral compound called DPT (deoxypodophyllotoxin, if you must know).
- Caution: If you're sure that you have found a juniper but aren't sure about the species, you can tell by tasting a tiny portion of the berry. If it's unpleasantly bitter, spit it out. However, you should avoid eating unidentified wild plants.
- Lookalikes: Some other juniper plants are toxic. Yew trees are toxic evergreens, but they don't produce berry-like cones.

Black Walnuts

Unlike the walnuts you're familiar with, black walnuts[93] take a little bit of work to get into. However, the bolder taste is more than worth the effort. You can harvest them from late October to early December. If harvested later, the nuts are easier to process and usually have a bolder flavor, but you risk getting beaten to the punch by squirrels.

- Common Name: Black walnut or eastern black walnut.
- Scientific Name: *Juglans nigra.*
- Habitat: You can find it in warm areas where there's plenty of light. It prefers fertile, wet, lowland soil. Look for it near bodies of water.
- Geographic Location: Native to eastern North America. Some black walnut trees grow in Europe.
- Distinguishing Features: The tree can reach over 100 feet high. It has distinctive dark, heavily ridged bark. The large, pointed leaves grow in groups, and each one is irregularly toothed and hairy underneath. The tree and leaves have a spicy smell. In early summer, the flowers form as catkins that turn into brownish-green plum-like fruits. The nuts grow inside. In fall, the fruits will fall, and the green hull will slowly turn black.

---

[93] (*Black Walnut (Juglans Nigra): Benefits, Supplements, and Safety*, 2019), (Shaw, 2010), (*Black Walnut (Juglans Nigra) - British Trees*, n.d.)

- How to Harvest: You can harvest the green fruit from the tree, but it's easier to just look on the floor for the fruit. By winter, they will have mostly turned black. But whatever you do, wear gloves and clothes you don't care about when harvesting and processing black walnuts. The juice stains horribly. The walnuts will need hulling, then shelling.
- Edible Part/s: The kernel or nut part of the fruit.
- Nutritional Information: High in protein, very high in manganese. Black walnuts also contain healthy fats and other vitamins and minerals.
- Caution: Black walnuts are infamously tricky to process. They are slippery when you hull them, and the black juice stains skin and clothing. The shells are also tough, so you'll have to use a hammer or a specialized nut cracker.

White Pine

There are around 80 species of pine growing across North America and even more around the world.[94] Some of these species aren't safe to forage, but the eastern white pine is. While it seems desperate to eat the needles, pine can be a tasty and valuable ingredient. In the worst-case scenario, it can make a survival situation in winter more plausible.

- Common Name: White pine or eastern white pine.
- Scientific Name: *Pinus strobus.*
- Habitat: In the wild, you can find white pine in upland forests. It's sometimes planted ornamentally in parks and large backyards.
- Geographic Location: This species is native to eastern North America. You can find it from Quebec as far south as Alabama and as far west as Manitoba.
- Distinguishing Features: White pines can grow up to 130 feet high. It has thick, dark gray bark with scaly plates or long ridges. The needles grow in bundles of 5 and are about 2 to 4 inches long. They're soft and feathery and remain on the tree for two years before being shed. The seed cones are cylindrical and slightly curved in shape and are usually 3 to 8 inches long.

---

[94] (*White Pine Vinegar — Four Season Foraging,* 2018)

- How to Harvest: When harvesting, bear in mind that older pine needles are healthier than young needles, but they do tend to be more bitter. They're easy to pick, as pine needles are less likely to stab you. Use your hands or a cutting tool to snip off the needles. Don't pick all the new growth from one spot of a single tree, but move around. Also, don't pick the leading tip (which is usually too high up anyway) so that you don't stunt the tree's growth.
- Edible Part/s: The needles are edible but must be cooked or prepared to be palatable.
- Nutritional Information: Pine needles are very high in vitamin C. Older pine needles have three and a half times as much vitamin C as oranges, making them a fantastic immunity booster.
- Lookalikes: Some pine species are toxic. Other needled evergreen trees can be confused with pine (like the spruce tree). If in doubt, check the needles. They should match the description above, as other evergreen trees have different-shaped needles.

How to Use Winter Forage

Most winter forage is best when cooked or processed as an ingredient. Even cranberries are tart enough that most people prefer to use them rather than eat them raw. Black walnuts are an exception, although you could argue that the hulling and shelling process is involved enough for anyone. You can use your winter produce to create some truly unique and powerful flavors.

Using Cranberries

Cranberries and highbush cranberries can be frozen, dehydrated, or used in sauces and jellies. They're high in pectin, and their tartness can cut through cheese and meat alike. Here's a cranberry sauce recipe that's quick, simple, and genuinely delicious.[95]

Cranberry Sauce

Ingredients:

- 4 cups of cranberries, fresh or frozen

[95] (Bauer, n.d.)

- 1 cup of sugar
- 1 orange
- 1 cup of water (or less)
- Salt

Method:

1. Rinse the cranberries and discard any debris or damaged berries.
2. Zest and juice the orange. Combine it with the water to make up 1 cup of liquid.
3. Put the water, sugar, and orange juice (with or without the zest) into a saucepan and bring to a boil. Stir to dissolve the sugar.
4. Add the cranberries and return to the boil. Lower the heat and cook until the cranberries have burst and the sauce has thickened slightly. Add some salt.
5. Let the sauce cool. It will continue to thicken. Store in a jar and keep in the refrigerator.

Notes:

- You can omit the orange and just use a cup of water, but it works very well with cranberries. You can also spruce up the sauce by adding spices like nutmeg or cinnamon.

Using White Pine Needles

Pine needles aren't anyone's first port of call for a meal, but you can use them to infuse things with flavor and that wonderful vitamin C. I'll show you how to make pine-infused vinegar and give you some ideas on how to use it.

Pine-Infused Vinegar

Ingredients:

- 1 cup of white pine needles, washed, dried, and roughly chopped
- 1 cup of white wine or apple cider vinegar

Method:

1. Prepare the white pine needles and put them into a sterilized glass jar.
2. Pour in the vinegar. It should completely cover the needles.
3. Seal the jar with a vinegar-proof lid and store in a cool, dark place for 4 to 6 weeks.
4. Once it's ready and you're happy with the flavor, strain into a different sterilized jar. Use within 6 to 12 months.

Notes:

- You can use this vinegar in salad dressings, marinades, and sauces.
- You can drink the vinegar medicinally, either straight or in a drink. Try adding ¼ cup of the vinegar to 12 fluid ounces of ginger beer for a delicious and healthy drink. You can even add it to cocktails.

To Conclude

While winter isn't quite as productive as other seasons, there's always something to find out in the wild. It just takes a little imagination. However, one wild edible that crops up in most seasons is the humble mushroom. Next, we'll look at some of the wonderful mushrooms you can find.

10

Foraging Edible Mushrooms

Mushrooms are a popular foraging choice for a few reasons. First, mushrooms are delicious. They're versatile, easy to cook, and wild mushrooms have unique flavor profiles that you can't find in stores. Mushrooms can also flourish all year round, with some even surviving through winter. Finally, mushrooms have a unique nutritional composition that makes them a great part of any balanced diet.

However, you have to be especially careful when identifying mushrooms. Failing to distinguish between a delicious edible mushroom and a poisonous mushroom can lead to severe harm or even death.

Mushroom Profiles

Mushrooms are categorized differently from plants, so these are mushroom profiles instead. But they do grow in the ground, and you're free to pick them to use in your cooking. Here are some profiles of my favorite edible mushrooms.[96]

There are hundreds of edible mushrooms that you can find in North America alone, let alone worldwide. However, these options are delicious, common, and, most importantly, easy to identify.

Please see the QR code for photographs of the following mushrooms.

---

[96] (*11 Edible Mushrooms in the US (And How to Tell They're Not Toxic)*, 2018), (Neuharth & Sammak, 2020)

Morel Mushrooms

For a brief period in spring, budding mushroom hunters get the opportunity to find one of the most prestigious wild mushrooms out there. People will walk miles for these delicious and odd-looking treats, and once you find a local grove, you'll find out why.[97]

- Common Name: Morel mushrooms.
- Scientific Name: *Morchella esculenta.*
- Habitat: Morels prefer moist soil and a lot of sun. You can find them in river bottoms, sparse forests, and burn areas. They also thrive in disturbed areas. Morels can be commonly found near elm and ash trees, as well as poplars, sycamores, and cottonwoods.
- Geographic Location: Found in North America and Europe. You can find some species worldwide.
- Season: March to September, depending on the climate. Your best bet is to hunt in April and May. In hotter areas, look in March, and in colder areas, look later in the season. The morel season can range from a few days to a whole month.
- Distinguishing Features: Morels are easy to identify by their unique honeycomb caps. The cap is usually brown or gray, and the stem is typically cream in color. The cap is also taller than it is wide, like an egg. Morels are small and completely hollow when you cut them in half.
- How to Harvest: Use a sharp cutting tool to cut the stem just above the ground.
- Edible Part/s: The whole mushroom is edible when cooked. Don’t eat morels raw.
- Caution: If the morel has an odd hole in the top, insects may have infested it.
- Lookalikes: There are several species known as false morels, some of which are toxic. Many of them have a brain-like structure rather than a honeycomb. If you cut a false morel in half, it won't be completely hollow.

---

[97] (Neuharth & Sammak, 2019)

Shaggy Mane Mushrooms

These distinctive mushrooms are very delicate, which means they only have a short shelf life. If you forage them, eat them within a day or so to get the best result. Believe me; it's worth it.

- Common Name: Shaggy mane mushroom, Lawyer's wig, or shaggy inkcap.
- Scientific Name: *Coprinus comatus.*
- Habitat: Grasslands and meadowy areas.
- Geographic Location: Found across North America and Europe.
- Season: June to November, depending on the climate and temperature.
- Distinguishing Features: This species is easy to identify because of its distinctive pale shaggy cap, which is covered in loose scales. As one of the names suggests, it looks a bit like a British barrister's wig. It's only edible before the gills turn black, so look out for white or pink gills.
- How to Harvest: Check the gills and, if they're white or pink, use a sharp knife to harvest them.
- Edible Part/s: The whole mushroom is edible cooked, as long as it doesn’t have black gills.
- Caution: As mentioned once or twice, mature shaggy mane mushrooms are inedible. Thankfully, the gills give it away.
- Lookalikes: It can be mistaken for the common ink cap, which is darker and doesn't have the shaggy scales. The common ink cap is only toxic if you consume alcohol within a day of eating it. If you eat common ink cap with alcohol, expect some unpleasant symptoms, including vomiting and diarrhea.

Chicken of the Woods Mushrooms

Bright and colorful, like a rooster's orange comb, you can spot these mushrooms from a mile off. You can sometimes find them in massive clusters. They have a rich, meaty flavor and are pretty sturdy. This makes them a fantastic chicken or tofu substitute in recipes.

- Common Name: Chicken of the woods mushrooms or sulfur mushrooms.

- Scientific Name: *Laetiporus sulphureus.*
- Habitat: Grows on freshly dead or mature hardwood trees, such as oak trees. You can sometimes find them on conifers. Look in forests.
- Geographic Location: Found throughout North America, more commonly east of the Rocky Mountains.
- Season: April to November, depending on the climate.
- Distinguishing Features: These mushrooms don't have gills and grow in large, fleshy clumps along hardwood trees, mainly oak trees. They're usually bright orange or sometimes salmon-pink on top.
- How to Harvest: Cut the young caps off, ignoring the tough stem or older caps.
- Edible Part/s: The young, fleshy cap is best to eat. You should cook this mushroom.
- Caution: If you can't identify the tree this mushroom is growing on, steer clear. If it's growing on a toxic or eucalyptus tree, you may be dealing with an inedible variety. If in doubt, stick to oak trees. The stems and older caps are unpleasant to eat and may cause stomach issues.
- Lookalikes: A similar mushroom known as hen of the woods is also edible. It is usually dark gray or cream-colored and grows in a rosette shape. They can be found in the same habitat as chicken of the woods and have a similar texture.

Chanterelle Mushrooms

Chanterelles are at least as popular and well-known as morels, and for good reason. They have a delicate peppery flavor and a richness that makes them deliciously decadent when cooked with lashings of butter. This is a true treat.

- Common Name: Chanterelles.
- Scientific Name: *Cantharellus cibarius.*
- Habitat: Commonly found in coniferous forests, as well as grasslands, birch forests, and beech forests. It largely depends on the location and the variety.
- Geographic Location: North America, Central America, Eurasia, and Africa.
- Season: Late summer to December, depending on the area.

- Distinguishing Features: These mushrooms are yellow, or gold in color and have a distinctive trumpet or funnel shape. They have gill-like ridges on the underside that run down the stem. Some people say that they have a fruity or woody smell.
- How to Harvest: As usual, use a sharp tool to cut the mushroom close to the ground.
- Edible Part/s: The whole mushroom is edible.
- Lookalikes: Chanterelles have a few lookalikes. Black trumpet mushrooms are, as the name suggests, black or dark gray and have a similar trumpet shape but no ridges. Black trumpets are edible and usually found in shady or damp areas. False chanterelles are more orange in color and have a darker center that lightens at the edges. It isn't dangerous, but it tastes bad and can upset your stomach. Jack-o-lantern mushrooms are toxic and have a similar color to chanterelles but have thinner and more pronounced gills.

Giant Puffball Mushrooms

This mushroom looks exactly as it sounds, making it really easy to spot. They can grow from between 4 to 20 inches across, although some reports suggest they can reach nearly 60 inches. Puffballs store well and are famously great in breakfasts.

- Common Name: Giant puffballs.
- Scientific Name: *Calvatia gigantea*
- Habitat: Grows in fields, meadows, and forests. They like disturbed areas.
- Geographic Location: Grows in North America and Europe. Prefers temperate climates.
- Season: Spring to fall.
- Distinguishing Features: Puffballs are giant, white, and rounded, with no identifiable stem, cap, gills, or gaps. They have solid, white flesh inside, so cut them open before cooking to ensure you have the right mushroom.
- How to Harvest: Pull or use a knife to cut away from the ground.
- Edible Part/s: The white flesh is solid.

- Caution: Mature puffballs are toxic, but you can tell the difference by cutting them open. Edible puffballs have white flesh, while older puffballs have greenish-brown flesh.
- Lookalikes: Some immature varieties of other mushrooms look a bit like puffballs. These may be toxic. If, when you cut it open, you find a differently colored interior, gaps, or the silhouette of a cap or gills, then discard it. It should be completely solid and homogenous.

Bolete Mushrooms

There are many different species of bolete, most of which are edible and delicious. One of the best of the bunch is the king bolete[98], which is considered a particularly fantastic mushroom. It's terrific in risotto dishes and has a robust flavor that can stand up to anything.

- Common Name: King bolete, porcini, or penny bun mushroom.
- Scientific Name: *Boletus edulis.*
- Habitat: Usually found in clearings of broad-leaved and coniferous forests. You'll only ever see them near trees.
- Geographic Location: Common in Eurasia, especially in Britain and Ireland. It's also found in North America and has been introduced to southern Africa and Oceania.
- Season: Can be found from spring to fall, but most commonly late summer and early fall.
- Distinguishing Features: Bolete mushrooms are usually tan or brown. They have a bulbous stem and a large cap with a light margin. When you slice the mushroom, you'll see pale-yellow tubes inside that end in tiny white or yellow pores. The stem is creamy and has a faint net pattern on the surface.
- How to Harvest: If you can, forage a few days after a summer rain. Use a sharp knife to pick the mushroom close to the ground.
- Edible Part/s: The whole mushroom is edible.

---

[98] (*Boletus Edulis, Cep, Penny Bun Bolete Mushroom*, n.d.)

- Caution: Boletes are infamously popular with maggots and other creepy crawlies. If the gills are greenish-yellow, then it likely has a maggot infestation. Cut open each mushroom you harvest to check for bugs.
- Lookalikes: The bitter bolete is edible but very bitter. It has a darker stem and pinkish pores. It has no toxic close lookalikes; just avoid anything with red or pink pores, and you'll be fine.

Toxic Mushrooms

While there are plenty of delicious edible mushrooms out there, you'll also encounter many toxic mushrooms. Poisonous mushrooms can have different effects depending on the specific toxins in them.[99] Knowing the toxin can help diagnose, allowing you to get swift and accurate treatment. Here are the categories you may come across.

Amanitin

This is the nastiest type of mushroom toxicity. Mushrooms that contain amanitin are most likely to result in death. They include the infamous Death Cap mushroom and the no less deadly Destroying Angel. Unsurprisingly, if something has "death", "destroying", "deadly", or "poison" in its name, don't eat it.

The main issue with this toxin is that major symptoms don't develop until 24 hours later, by which time the body has completely absorbed it. The initial symptoms include:

- Nausea
- Vomiting
- Abdominal pain
- Diarrhea

You may experience more severe symptoms like:

- Violent vomiting
- Bloody diarrhea
- Acute or extreme abdominal cramping

---

[99] (Fischer, n.d.)

Some people then seem to improve for 24 hours, which gives the impression of minor poisoning or a stomach bug. Even if you've gone to a hospital, they may misdiagnose you and release you from medical care. After these 24 hours, an improperly diagnosed patient will suffer from:

- Liver failure
- Kidney failure
- Coma
- Permanent organ damage
- Death

If you've eaten a suspect mushroom, seek medical help, even if the symptoms seem mild. Take a sample of the mushroom with you so someone can identify it and so that you can receive potentially life-saving treatment.

Gastrointestinal Toxins

This is the most common type of poisoning you'll encounter and, thankfully, usually the least severe. Mushrooms may contain many different kinds of toxins and proteins that can disagree with you. Even a typically edible mushroom can affect certain people if improperly cooked or eaten in large quantities. Even if you've successfully identified a mushroom, only eat a small amount to start with.

Symptoms usually develop within an hour, or 4 hours at most, and include:

- Nausea
- Vomiting
- Diarrhea
- Abdominal cramping

While this type of poisoning is usually less dangerous, it can still be dangerous if the symptoms are severe. In rare cases, extreme vomiting and diarrhea have led to electrolyte depletion and dehydration, which may cause heart failure and death.

Muscarine

This toxin is unusual and exists in a few common mushrooms, such as Jack O'Lantern mushrooms and some species of the *Boletus* genus (look out for orange/red-colored pores and blue stains).

These mushrooms can cause the following symptoms:

- Excessive tear secretion
- Excessive perspiration
- Excessive salivation
- Difficulty breathing
- Low blood pressure
- Irregular heartbeat
- Nausea
- Vomiting
- Diarrhea

These symptoms occur within half an hour of consumption. In severe cases, they can result in death from respiratory failure. As usual, seek medical attention and bring a sample. This type of toxicity is usually treated with atropine.

Isoxazole Derivatives

Several poisonous mushrooms contain muscimol and ibotenic acid, such as the Fly Agaric. The symptoms are frightening but usually temporary and unlikely to result in death or permanent damage. This is fortunate because there is no effective treatment. The symptoms occur between 30 minutes and 2 hours after consumption. They include:

- Delirium
- Manic behavior
- Inebriation
- Perception issues (small objects appear large)
- Drowsiness

- Nausea
- Vomiting
- Delusions
- Convulsions
- Hallucinations

These symptoms may last up to 4 hours or even longer. Seek medical assistance, and don't leave someone in this condition alone.

Gyromitrin

Gyromitrin is a compound that, when heated, can produce monomethylhydrazine (MMH). MMH is a rocket fuel and, as you'd expect, is not great to eat. It can interfere with how your body handles vitamin B6, affecting metabolism. Symptoms may include:

- Bloating
- Abdominal pain
- Diarrhea
- Headaches
- Nausea and vomiting (after 7 hours)

This toxin can cause severe liver damage, resulting in death. MMH is also highly carcinogenic. It can be found in some toxic false morels and mushrooms in the Gyromitra genus.

Coprine

Mushrooms that contain the amino acid coprine are an interesting case. The mushroom itself isn't toxic, and you can enjoy it safely unless you've recently consumed (or will consume) alcohol. Coprine can interact with alcohol in the body, worsening the unpleasant symptoms of alcohol poisoning. Ink cap mushrooms are the most well-known type of mushroom that contains coprine.

Poisoning can occur if you consume alcohol within a few days before or after eating mushrooms containing coprine.

Psilocybin/Psilocin

The infamous "magic mushrooms" produce their effects because of these compounds. Mushrooms containing this toxin interact with the brain, causing symptoms including:

- Hallucinations
- Giddiness
- Anxiety
- Impaired perception
- Euphoria
- Drowsiness
- Muscle weakness

These symptoms typically progress over time, and after a couple of hours, most people will fall asleep and experience intense dreams before awakening. These symptoms can be dangerous as someone experiencing extreme hallucinations (or having a bad trip) may be unaware of their environment.

How to Forage Mushrooms

When foraging mushrooms, it's crucial to ensure that you've correctly identified whatever you're harvesting.[100] Never eat a mushroom if you can't reliably identify it. Here are some tips to aid you in mushroom identification:

- Use a field guide (or this book) that provides a mushroom profile of your intended forage.
- Stick to a few types of mushrooms that you can easily identify. This is especially important for novices.
- If possible, go mushroom foraging with a guide who knows the area and the types of mushrooms (both edible and toxic) that you can find.
- Don't be afraid to touch a mushroom. Even dangerously toxic mushrooms can only harm you when ingested. This can help you to get a closer look at it.

---

[100] (Brenner, 2018)

- Avoid mushrooms in poor condition, such as moldy or damaged ones.
- Don't harvest mushrooms that are too young or old. Even edible mushrooms may only be edible at certain life cycle stages. Learn these stages and how to identify your target mushrooms.

When you find a mushroom, carefully inspect all of its parts. Some mushrooms look very similar to others, especially in different stages in their lives. Even toadstools, which usually are easy to spot when mature, look completely different when young. Pick your chosen mushroom with a sharp knife and check:

- The shape of the body.
- The cap, namely the width, shape, and color.
- The underside of the cap. Does it have pores, spines, ridges, gills, or tubes? Is there a veil? Check the color, construction, and spacing of these features.
- Check the stem, look at the color, and any identifying features.
- Smell the mushroom.
- Identify the mushroom habitat and substrate. Does it grow on trees, leaves, soil, or wood chips?
- What season is it? Different mushrooms grow at different times.

Here are some more foraging tips[101]:

- Have a specific mushroom in mind so that you know what to look for.
- Be patient. Walk slowly and carefully. Watch the ground.
- Wear sensible foraging clothes.
- Bring a suitable container, such as a wicker basket. A plastic bag will speed up decomposition.
- Bring a torch and, ideally, a small magnifying glass.
- Bring a cutting tool to harvest the mushrooms without damaging them.
- Try a small portion of the mushroom before eating in large quantities.

---

[101] (Sayer, 2018), (*This Is How You Forage Mushrooms Safely (Anyone Can Do It) – Mushroom Grove*, n.d.)

How to Prepare Mushrooms

Once you've collected some mushrooms, it's time to get them ready to eat. You should cook most wild mushrooms before eating. A few species are edible raw, but it's usually safest (and tastiest) to cook them. How you treat your mushrooms may differ depending on the variety, so bear that in mind.

The first step is to clean your mushrooms.[102] The recommended method is to brush your mushrooms clean, but this can take ages. Believe it or not, you can use water to clean your mushrooms, as long as you don't soak them. Put them in a colander and spray them with water until the dirt washes away. The two major exceptions to this are puffballs and morels. Never get puffballs wet, as they absorb water very quickly. With morels, you will need to use running water and a brush to get the dirt and grit out of the honeycomb cap.

Next, it's time to cook them. Mushrooms go well in almost anything, although different mushrooms have different textures and flavor profiles. Some are meaty and firm, while others are delicate. They usually work well in stews and soups, but I'm a fan of sautéing my mushrooms in butter.

If you want perfectly cooked mushrooms that aren't soggy, then follow these steps:

1. Wash mushrooms and pat dry.
2. Heat a frying pan to medium-high heat.
3. Put the mushrooms into the pan and spread them evenly.
4. Leave the mushrooms alone, letting the steam escape. After a minute, add a tablespoon of butter and let them fry until golden brown. Enjoy with fried onions.

Recipes to Try

Here are a couple of great mushroom recipes that work brilliantly with wild foraged mushrooms.[103]

---

[102] (Overhiser, 2020), (*How to Cook Mushrooms - You Won't Believe the Secret!*, 2020)

[103] (Fishman & Jones, 2020), (Sjdin, n.d.)

Mushroom Risotto

Creamy, mushroomy, and delicious. Mushroom risotto is the ideal way to get the most out of your wild mushrooms. Try it yourself. This recipe serves 6.

Ingredients:

- 2 pounds of wild mushrooms, thinly sliced
- 2 medium shallots, diced
- 1 clove of garlic, finely diced
- 1½ cups of Arborio (risotto) rice
- 6 cups of chicken or vegetable stock, or as needed
- ½ cup of dry white wine
- 3 tablespoons of olive oil
- 4 tablespoons of butter
- ⅓ cup of freshly grated parmesan cheese
- Finely chopped chives, as garnish
- Salt and freshly ground pepper as seasoning

Method:

1. Prepare the vegetables and keep the broth warm in a saucepan over low heat.
2. Put 2 tablespoons of oil in a large saucepan over medium heat. Add the mushrooms and cook for 3 minutes or until tender. Set the mushrooms aside with their liquid in a bowl.
3. Add the remaining oil to the saucepan and cook the shallots until softened. Add the garlic, then the rice. Cool and stir until the rice is coated in oil and has taken on a pale, golden color. It should take 2 minutes.
4. Pour in the wine, constantly stirring until it's completely absorbed.
5. Add ½ cup of broth to the rice and stir until it's absorbed. While stirring, repeat this process until the rice is tender yet slightly firm (al dente) and the liquid is completely absorbed. It should take about 15 to 20 minutes.

6. Remove from the heat. Stir in the mushrooms and their liquid, along with the butter, chives, and cheese. Keep stirring until everything is melted, then season with salt and pepper and serve immediately.

Vegan Cream of Mushroom Soup

Creamy, comforting, and surprisingly vegan, this soup is ideal for cozy days in.

Ingredients:

Soup

- 12 ounces of wild mushrooms, roughly chopped
- 1 large onion, finely chopped
- 2 shallots, finely chopped
- 4 garlic cloves, thinly sliced
- ¼ cup of olive oil
- 4 garlic cloves, thinly sliced
- ⅓ cup of dry white wine
- ¼ cup of raw cashews
- 1 tablespoon of miso paste
- Salt and freshly ground black pepper

Optional garlic oil

- 3 tablespoons of extra-virgin olive oil
- 3 garlic cloves, thinly sliced
- 1 tablespoon of thyme leaves
- Salt and freshly cracked black pepper

Method:

1. Prepare the vegetables and heat the oil in a heavy saucepan or pot over medium-high heat.

2. Brown the mushrooms; it should take about 5 minutes or so. Once golden brown, use a slotted spoon to transfer the mushrooms to a bowl, leaving the oil behind.
3. Add the onion and shallots to the pot and turn down the heat. Cook until very soft and translucent, but don't let it brown. This should take 10 minutes. Add the garlic and cook for another few minutes, or until soft and fragrant.
4. Add the wine and cook until almost evaporated. Add 5 cups of water and the mushrooms to the pot and bring to a simmer.
5. Transfer 2 cups of the soup to a blender and add the cashews and miso. Puree until smooth, then return to the pot. Simmer for 15 minutes, then add salt and pepper.
6. A few minutes before serving, put the garlic oil ingredients into a small saucepan and simmer gently until the garlic is just beginning to turn gold and smells delicious.
7. Ladle the soup into bowls and drizzle with the garlic oil. Serve with crusty bread.

To Conclude

Mushrooms are a fantastic thing to forage, as they're always delicious and versatile. However, it can be tricky to avoid harmful mushrooms, especially as a novice. If you're careful, forage with a guide, and stick to what you know, you should get the hang of it. In the next chapter, we'll have a brief look at how to preserve and store wild edibles to truly get the most out of them.

11

Preserving and Storing Wild Edibles

One thing that you're sure to have noticed about foraging is that it depends heavily on the season. But it is possible to enjoy the fruits of your labor all year long once you unlock the secrets of preservation and storage.

Some preservation methods enable you to store foods for years, but I recommend that you eat your preserved food within a year. Keeping them longer can negatively affect their flavor, texture, and nutritional value. Still, by the time that year is over, you can forage your favorite wild edible foods all over again.

So, why is it so crucial for a forager to learn how to preserve and store food safely? The key word there is "safely". Improper preservation and storage can lead to food poisoning. It would be a shame if you spent all that time ensuring that something is edible, only to be poisoned by spoiled food. The good news is that proper preservation techniques allow you to:

- Keep your foraged finds edible for longer
- Maintain flavor and freshness
- Maintain nutritional value of foraged food
- Create new and exciting flavors from your foraged food

Without any further ado, let's look into how you can preserve your food.

Preparing Food for Storage

As soon as you forage plants or mushrooms, they will start to decay. Whether you plan to preserve them long-term or use them within a few days, it's vital to ensure that your foraged edibles stay fresher for longer. You need to prepare them and ensure that they are stored until you use them. These steps are applicable for everything you forage[104]:

1. When collecting wild edibles, ensure that you don't harvest damaged or overripe specimens. Be careful so that you don't crush delicate foraged food, like berries. Brush off excess dirt, debris, and any insects in the field.
2. Put things in containers or baskets so that they don't get crushed. Ideally, different types of forage (roots, leafy greens, berries, mushrooms, etc.) should be kept in separate containers. If this isn't possible, make sure that the heavier items are at the bottom.
3. When you get home, wash your hands and change your clothes, if you need to.
4. You will need to clean your foraged food quickly when you get back. Methods can differ slightly depending on what type of food it is. Mushrooms should be brushed clean or rinsed and patted dry. Keep them in baskets or plastic bags to avoid moist conditions. Berries can be soaked for a while to remove any debris or dirt, then rinsed in a colander. Greens should be rinsed and patted dry, as you would with lettuce and spinach.
5. Once everything is clean, you can either use it, store it in the refrigerator, or process it for preservation.

You also need to consider how certain factors affect spoilage when it comes to proper storage and preservation. Here's a quick overview of what to keep an eye on:

- Moisture/Humidity. Food will spoil more quickly if an environment is moist, as most microbes prefer damp areas.
- Temperature. Most microbes thrive in warm temperatures. The colder something is, the slower it will spoil. In winter, you might be able to keep things safe in a cupboard. Foraged fruits and vegetables may degrade within days or even hours in summer.

---

[104] (*Harvesting and Processing Edible Wild Plants*, n.d.)

- Light.[105] Even light can cause food to degrade. Light can increase the temperature of food, discolor food, and even lessen your food's flavor and nutritional value.

When you consider these factors, it's clear that keeping your food in a cool, dark, and dry environment is the best way to hold off spoilage. A refrigerator is typically cool and dark, and as long as you don't keep your food in a plastic bag, it will stay relatively dry. This applies whether you plan to eat or cook your foraged food fresh or store it after being processed for preservation.

Preservation Methods

Long-term storage requires processing food using some kind of preservation method.[106] When preserving food, it's essential to make sure that you process them early and make sure that you use the best specimens. A common mistake is only to preserve foraged food when it's about to go off, and you're panicking because you don't want it to go to waste. However, you will always end up with an inferior product, which may still spoil.

Some standard methods include:

- Drying/dehydration. The lack of moisture means that microbes can't thrive.
- Freezing. Freezing drastically slows spoilage by killing microbes or causing them to become dormant.
- Canning. The food is cooked, killing any microbes. Then it's kept in an airtight, pressurized environment that microbes can't access.
- Pickling. Highly acidic environments are inhospitable to microbes.
- Fermenting. Usually, the food is put in an environment that is only hospitable to beneficial microbes. For example, lactobacillus enjoys salty environments. Good microbes push out harmful microbes, some of which also create an acidic environment.
- Preserving in alcohol. Alcohol pulls out water and prevents the growth of microbes. The alcohol and the food also infuse each other with flavor.

---

[105] (Lee, 2017)

[106] (Soken, 2020), (*Sugar as a Preservative*, 2021), (Raab, n.d.)

- Preserving in sugar. Sugar absorbs moisture and can create an airtight environment. Honey is an excellent preservative.
- Preserving in oil. Oil creates an airtight environment, preventing bacterial growth. However, it's usually best to keep food maintained this way in the refrigerator as well.

Except for freezing, these preservation methods also affect the taste and texture of your food. Dehydrated foods have a more concentrated flavor. Even when rehydrated, formerly dried food will have a different texture and a slightly different nutrient profile. The other methods don't just keep your food going for longer, but they can add a whole new dimension to it.

As you'd expect, different types of foraged food work best with different preservation methods. Here are the best ways to preserve your wild edible forage.

Preserving Foraged Mushrooms

Once your mushrooms are clean and trimmed (if necessary), it's time to get preserving. Mushrooms can get slimy quickly, especially if kept in plastic bags. They can also shrivel up, which isn't as unpleasant, but it still means they aren't at their best. Before preserving anything, make sure that the mushrooms are clean and in good condition. It may be an idea to trim the fibrous stem root and check underneath the cap. Once the mushrooms are perfect, you can get started.[107]

Freezing

Because mushrooms are so full of moisture, you can't just throw them into the freezer and call it a day. That water will crystallize, which affects the texture of your mushrooms. Sometimes the mushroom is okay after thawing, but it can over-soften into a mush or even end up rubbery. Follow these steps for the best result:

1. Cook your mushrooms. You can boil, fry, or bake them in the oven. Cooking reduces enzyme activity and kills any bacteria in the mushroom.
2. Pat down the mushrooms and remove as much moisture as possible. A salad spinner can work wonders, but paper towels will do fine.

---

[107] (Lee, 2022)

3. Put the mushrooms in a freezer bag and arrange them flat in a single layer.
4. Then, remove as much oxygen as possible from the bag. You can do this with a vacuum sealer or by manually squeezing the air out of the bag. This prevents ice from forming and saves space in the freezer. Seal the bag.
5. Label the freezer bag with the date and the type of mushrooms. They should last for six months or up to a year if vacuum sealed.

Drying

Another way to preserve mushrooms is to get rid of all that moisture. You can buy dehydrated mushrooms in the stores, and they're usually relatively expensive. That's because most dehydrated mushrooms are an expensive variety to start with (like porcini or shitake), and the flavors are so concentrated. But you can dry delicious wild mushrooms at home. You can use a dehydrator, drying cabinets, or simply air dry mushrooms. Once dried, store in a cool, dark place.

Air-drying mushrooms doesn't require any special equipment. However, it isn't as thorough as other methods.

1. Cut the mushrooms into slices.
2. Cover a wire tray with baking paper.
3. Evenly arrange the mushrooms on the tray without overlapping.
4. Put the tray into an airy, sunlit room for at least 48 hours. Mushrooms can absorb surrounding smells, so bear this in mind.

Drying cabinets are a smell-free solution, allowing you to confine your mushrooms. If you have a spare cabinet with trays, you can put a low heat lamp at the base of the cabinet and repeat the air-drying process there. Just leave it overnight for dried mushrooms.

The quickest and most reliable way to dry mushrooms is to use an electric dehydrator.

1. Slice the mushrooms and arrange them in single layers on the dehydrator trays.
2. Set the dehydrator to 110-120ºF and dry for 6-10 hours. Check on them periodically. They should be brittle and crisp.

Canning

Canning will retain flavor and water alike, so the mushrooms are preserved in their most natural state. However, canning isn't as foolproof as the other methods. If done incorrectly, it can result in food poisoning. I'd recommend only canning certain types of mushrooms, namely button or champignon mushrooms. Canning wild mushrooms are riskier. Here are the basic steps:[108]

1. Prep the mushrooms, trim off the stems, wash, and cut them to equal sizes.
2. Cook the mushrooms in a pot of boiling water for about 5 minutes.
3. Once you cook the mushrooms, pour them into the jar or can, along with the boiling liquid. Leave at least 1 inch of headspace at the top of the can.
4. Add a ¼ teaspoon of salt per pint of liquid. You can also put in a little bit of lemon juice to help with color retention, as well as other flavorings.
5. Check the jars or cans for damage or abnormalities.
6. Put the jars or cans into your pressure canner and use them according to the manufacturer's instructions. Process for 45 minutes.
7. Allow the canner to depressurize before opening, then remove the jars and let them cool. They should seal as they cool. Check the seals the next day, then label and store them.

Preserving Foraged Herbs

Fresh herbs are delicate, which means they can degrade very quickly. Act quickly and freeze or dry them to keep those delicious foraged flavors going.[109] As with other wild edibles, wash your herbs before you start and choose the best specimens.

Freezing

There are a few methods for freezing foraged herbs, which give you the taste of fresh herbs.

Freezing whole herbs individually allows you to use them in anything.

---

[108] (Miller, 2020)

[109] (*How To Preserve Fresh Herbs: Tips & Tricks for Storing*, 2021)

1. Bring a pot of water to a boil and set aside a bowl of ice water.
2. Cut the herbs into the boiling water for 15 seconds, then plunge them into the ice water. Leave them there for 1 minute.
3. Pat the herbs dry, then arrange them in a single layer on a baking sheet.
4. Put the sheet in the freezer for 1 hour, then transfer the now-frozen herbs to a plastic bag. Return to the freezer.

Freezing herbs in water and oil is a great way to keep them fresh. Water is best for herbs to be used in soups and stews, while olive oil adds another dimension for flavor.

1. Chop the herbs into small pieces and put them in the compartments of an ice cube tray.
2. Fill the tray with either water or oil.
3. Place in the freezer for 24 hours. Once completely frozen, transfer to a labeled plastic bag and return to the freezer.

Drying

You can air-dry herbs quite successfully or use the oven or dehydrator to get thorough results more quickly.

You'll need kitchen string, scissors, and clothes pegs to air-dry your herbs.

1. Group your herbs into bundles and keep the stems together at the bottom.
2. Wrap them tightly with string and tie them in a knot. Snip the string.
3. Hang the bundles upside down in a completely dark and dry area.
4. After 5-10 days, they should be completely dry. Store them as they are in airtight jars and crumble them only when used for the best results.

Drying with an oven or dehydrator has similar methods.

1. Preheat the oven to 200ºF or the dehydrator to 95-115ºF.
2. Arrange the herbs in a single layer on a tray and put them in the oven/dehydrator.
3. In the oven, they will take 2-4 hours to dry. The dehydrator will take 1-4 hours. Check the herbs often; they're ready when they're brittle.

4. Store in a cool, dark place.

Preserving Foraged Vegetables

Seasonal vegetables are delicious, and preserving them means that you can continue to enjoy their unique flavors at their best.[110] As always, clean and dry the vegetables first and make sure you have good specimens.

Blanching

Blanching is a recommended first step before freezing or preserving many vegetables, foraged or otherwise. Essentially, you're shocking the vegetable with sudden heat and then cold. Blanched food keeps its color better and, in some cases, has a better texture.

1. Bring a pot of water to a boil and prepare a bowl of ice water.
2. Put the prepared vegetables into the boiling water. The time may vary depending on the vegetable, but you should only blanch even fibrous vegetables like parsnips or asparagus for a few minutes. You should only blanch other vegetables for 15 seconds or so.
3. Put the vegetables in the bowl of ice water.

If you're dealing with soft greens and don't plan to freeze or dehydrate them, it's still worth soaking them in cold water to refresh them and extend their shelf life.

Freezing

When you freeze home vegetables in a home freezer, the moisture inside them expands. This can change the texture, meaning that the vegetables break apart more quickly when cooked. They're delicious raw in salads, though.

1. Once you've blanched the vegetables, vacuum seal them, store them in a freezer bag, or lay them on a metal tray to freeze in a single layer. With the last method, you can put them into a freezer bag once frozen, and they will separate easily.

---

[110] (Logiste, 2015), (*Wild Food: How to Forage and Store Wild Greens*, n.d.)

Drying

As usual, you can air-dry vegetables or dry them in a dehydrator. The times differ with different vegetables, but the basic method is the same.

First, pat your vegetables dry and cut them into pieces if you haven't already and it's necessary. It will speed up the process.

If air-drying, tie the stems together with string to create a bundle, then hang them in a warm, dry, and airy area. Ideally, it should be dark. If you can't tie the vegetables in a bundle, you can arrange them on a wire rack and put them in the same area.

If using a dehydrator or oven, look at the recommended times and temperatures. As a general rule, you should dry vegetables at 125°F. If you aren't sure of the time, just check after every hour until it's satisfactory.

Preserving Foraged Fruits and Berries

Foraged fruits and berries can be frozen and dehydrated using the methods outlined above (dry at 135°F). However, I thought we'd look at some preservation methods that create new flavors and allow you to put your forage to good use.[111]

First, if you want to keep your berries fresh in the fridge, you should get rid of any trace of mold. You can do this by giving them a 2-minute white vinegar bath, then thoroughly rinsing away the vinegar and any dirt. Then, store your berries in a clean, ventilated box lined with a paper towel.

Canning

When canning fruits and berries, it's best to use the water bath method. This will give it a shelf life of around one year. You should ensure that the fruits and berries are clean and cut (if applicable) into equal shapes and sizes. If it has a skin or stones/seeds, get rid of them.

1. Sanitize the canning jars, lids, and bands.
2. Fill a large pot with water and put a metal rack inside. Place the jars into the pot. Bring the water to a simmer for 10 minutes.

[111] (Greaves, 2021), (Parker, 2020), (Russell, n.d.)

3. Meanwhile, simmer the fruit with a canning liquid for 2 minutes. You can use sugar syrup, fruit juice, or even water.
4. Put the fruit and liquid into jars, leaving an inch of space. Pop any air bubbles using a spatula or tapping the jar on the counter.
5. Wipe and dry the rims, then put the lids and bands on the filled jars. Tighten until you start to feel resistance, don't go any further.
6. Put the jars back into the pot; the water should be a couple of inches above the jars. Process for 10 to 15 minutes, then let cool for another 15 minutes before removing the jars.
7. Let them cool for 12 hours. Check for a seal. If it's sealed, then store in the cupboard. If not, refrigerate.

Pickling

You can pickle with vinegar or with a salt solution. Using a 2-3% salt solution causes the fruit to ferment, creating lactic acid. The lactic acid then pickles the fruit. You can also use these methods for vegetables. I'll cover the vinegar method here.

1. Create a pickle brine. A simple ratio is 1 cup of vinegar, one cup of water, ½ a cup of sugar, and a tablespoon of salt.
2. Bring the brine to a boil.
3. Meanwhile, pack the fruit into a clean jar along with any whole spices. A cinnamon stick works a treat.
4. Pour the boiling brine into the jar. Seal and store in the fridge.

Jam

You can make a straightforward jam using your fresh berries.[112] The one thing to remember is that jams set with pectin. Some fruit and berries have a lot of natural pectin, like apples and cranberries. This is what the lemon juice is for, although jam sugar or a chopped apple also help.

---

[112] (Guide, n.d.)

1. Chop 3 pounds worth of fruit into 1-inch pieces.
2. Put the fruit, 1½ pounds of sugar, and a pinch of salt into a heavy pan.
3. Bring the mixture to a boil, crushing the fruit with a potato masher to release more liquid.
4. Stir until the sugar has dissolved, then add 2 tablespoons of lemon juice.
5. Keep stirring for 10-12 minutes until the mixture is thick and clumpy.
6. Skim off any foam and ladle the jam into clean jars. Let cool, then refrigerate.

Jelly

Unlike jam, fruit jelly is smooth and perfectly homogenous. While I personally like the rustic texture of jam, jelly is easy to make and a great way to eat more fruit.[113]

1. Put 1 pound of chopped and destemmed fruit/berries into a food processor and blitz into a pulp.
2. Put the pulp into a pot along with ¼ of a cup of water and bring to a boil. Simmer for 10 minutes.
3. Strain the pulpy juice using a cheesecloth. Discard the pulp and put the juice into a pot.
4. Stir 1 ounce of powdered pectin into the juice, then bring to a roiling boil. Add 2 cups of sugar and a pinch of salt. Boil for another minute, then take off the heat.
5. Skim off any foam and transfer the hot jelly into clean jars. Let cool, then refrigerate.

Preserving Foraged Edible Flowers

Foraged edible flowers add a lovely floral note to fresh salads or as a garnish. Like herbs, they don't last very long when fresh. These preservation methods give the flowers a new lease of life and allow them to impart their flavor to other dishes.[114]

---

[113] (*How to Make Easy Homemade Jelly: Basic Jelly Recipe With Tips, Ingredients, and Flavors - 2022*, 2021)

[114] (Breyer, n.d.)

Vinegar

You can use flower vinegar in salad dressings and other recipes calling for vinegar. It adds a delicate but no less pleasant note.

- Put 2 cups of white wine vinegar in a clean bottle with ½ cup of flower petals.
- Store in a cool, dark place for a week. Strain flowers and use vinegar as you wish.

Honey

Different types of flowers produce different flavor combinations. Lavender and rosemary blooms have a strong flavor, while rose petals are floral. You can then use the honey as normal.

- Put 1 cup of flower petals into a reusable tea bag or muslin/cheesecloth bundle, then add to a pound of honey.
- Leave in a bright, sunny place for a week (a windowsill is perfect) and check the flavor.
- If you're happy with it, remove the bag. If you want a stronger flavor, leave it in for a bit longer.

Syrup

Flower syrup is an excellent addition to cocktails and can also be used to top sweet dishes.

- Boil 1 cup of water, 3 cups of sugar, and 1 cup of flowers together for 10 minutes.
- Strain, then store in the refrigerator for 2 weeks.

Sugar

Sugar can absorb moisture and flavor, while flower petals add color and texture. Use this sugar to finish off baked goods or to rim cocktail glasses.

- Stir 1 cup of chopped flower petals with 2 cups of granulated sugar.
- Let it sit for a week. Enjoy.

Alcohol

Create some fabulous floral cocktails with infused alcohol in glasses rimmed with flower sugar. Vodka works best due to its relatively neutral flavor.

- Add ½ cup of flower petals to 2 cups or so of vodka in a bottle.
- Let it sit for 48 hours, then strain.

To Conclude

It's always a shame to let things go to waste, which is why it's so important to learn how to preserve our food. You can also make some new and delicious products. However, you can use foraged food in other ways. The next chapter will include some new recipes to try out.

12

More Recipes for Foraged Wild Edibles

Whether fresh or preserved, wild edible plants and mushrooms are fantastic to cook with. They make great ingredients for a variety of dishes and provide unique flavors, brilliant health benefits, and, best of all, they're free. There's nothing as satisfying as creating a delicious meal from scratch using foraged ingredients. I've already covered some recipes, but here are some more to try.

Desserts and Snacks

Who doesn't like desserts and snacks?[115] These sweet treats can convert anyone to the delights of foraged food, even picky children. If you take them foraging with you, reward them with one of these recipes.

Dandelion Banana Bread

Easy to bake and even easier to eat, this banana bread recipe is the perfect way to use dandelion flowers.

Ingredients:

- 1 large ripe banana (overripe bananas are perfect)
- ½ cup of olive oil

---

[115] (*Wild Food Dessert & Snack Recipes*, n.d.),

- 1 egg
- ⅓ cup of brown sugar
- 1¼ cup of plain, unbleached flour
- ⅓ cup fresh dandelion flower petals
- 1 teaspoon baking powder
- ½ teaspoon baking soda
- Optional - ¼ cup of chopped nuts or chocolate chips

Method:

1. Preheat oven to 350°F.
2. Mash the banana into a paste, then add the oil, egg, and sugar. Mix well.
3. Stir in the rest of the ingredients and mix until well combined.
4. Pour the mixture into a greased loaf tin and bake for 20-25 minutes.

Pine Needle Cookies

This recipe gives you that vitamin C boost and a fresh, piney flavor in these delicious packages. I'll tell you how to make pine powder, which you can then use for the cookies.

Pine Powder:

1. Collect pine needles on the branches and hang them in your home to air-dry them.
2. After 2-3 weeks, remove the dried needles from the branches.
3. Put the needles into a spice or coffee grinder and grind into powder.

Ingredients:

- 3 cups of unbleached plain flour
- 1½ cups of cane sugar
- 8 tablespoons of pine powder
- ½ cup of melted butter
- 3 eggs

- 1 teaspoon of vanilla

Method:

1. Preheat the oven to 325°F. Put the flour, sugar, and pine powder into a bowl.
2. In another bowl, mix the butter, eggs, and vanilla. Add to the dry ingredients and combine.
3. Roll the dough into small balls (¾ the size of a golf ball) and put onto a dry baking sheet. Flatten the cookies with a fork until they're about ¼-inch thick.
4. Bake for 10-12 minutes. Let cool, then enjoy.

Salads

The best way to use fresh foraged greens and flowers is to include them in a seasonal salad,[116] especially if the leaves are young and sweet. You can also put fruit and nuts in salads, to make them more interesting.

Chickweed Salad

Chickweeds are delicious when they're fresh, and this zesty salad is quick and straightforward to put together.

Ingredients:

- 1 cup of chickweed leaves, washed and drained
- 1 bunch of scallions or 1 small red onion
- 1 raw beet
- Pinch of salt
- 3 tablespoons of coconut oil or extra-virgin olive oil
- 2 tablespoons of vinegar (white wine, red wine, or apple cider vinegar)
- ½ teaspoon of mustard (dijon or whole grain)

Method:

[116] (*Wild Food Salad & Sautée Recipes*, n.d.)

1. Thinly slice the scallions or red onion. Grate the beet.
2. Put into a bowl with the chickweed leaves.
3. Combine the oil, vinegar, and mustard to create a dressing. I like to put it into a small Tupperware box with a lid and shake it, so it emulsifies.
4. Toss the dressing through the salad, then sprinkle with salt.

Plantain Salad

This nutritious summer salad is bright and refreshing, a perfect addition to a barbecue or a light lunch.

Ingredients:

- 2 cups of plantain leaves, finely chopped
- ½ cup of cabbage, finely chopped
- 1 large (28 ounces) can of chickpeas/garbanzo beans, drained
- 1 onion, finely chopped
- 1 celery stalk, finely chopped
- 1 large garlic clove, finely chopped
- 1 teaspoon of salt
- ⅛ cup of white wine vinegar
- ⅛ cup of olive oil

Method:

1. Prepare the ingredients.
2. Put everything except the oil and vinegar into a large bowl and refrigerate until chilled.
3. Once chilled, add the oil and vinegar and mix. If it's a little dry, add more liquid.

Soups

A great way to use fresh and preserved foraged edibles alike, soups[117] can be comforting in fall and winter and a delicious quick meal in spring and summer.

Lamb's Quarters Soup

The soup has a unique, zesty flavor, and it's perfect for a hot summer's day.

Ingredients:

- 2 plum tomatoes, cut in half
- 3 cups of water
- 3 cloves of garlic
- ½ of a lime, juiced
- 1 tablespoon of olive oil
- 1 teaspoon of honey
- ½ teaspoon of sea salt
- 1 tablespoon of butter or coconut oil
- 2 cups lamb's quarters leaves, finely chopped
- ½ of an avocado, finely chopped
- 1 red pepper, thinly sliced
- 1 large onion, finely chopped
- 2 celery stalks, very thinly sliced
- Optional - Edible flowers for garnish

Method:

1. Prepare the vegetables.

---

[117] (*Burdock Root and Miso Soup Recipe*, n.d.)

2. Put the tomatoes, water, garlic, lime juice, olive oil, honey, and salt into a food processor and blend until smooth.
3. Put the butter/oil into a saucepan and sauté the red pepper, onion, and celery until soft.
4. Add the blended liquid to the saucepan, then add the lamb's quarters and avocado.
5. Simmer on low heat (don't boil) for 20-40 minutes to let the flavors combine.
6. Garnish with edible flower petals and serve.

Stinging Nettle Soup

This is a great way to get all of the nutrients that stinging nettles have to offer and provide a refreshing meal.

Ingredients:

- 4 cups of young nettle leaves
- 4 cups of spinach
- 4 cups of chicken or vegetable stock, hot
- Cold milk
- 4 cold cooked sausage (vegetarian sausages will work)
- 3 tablespoons of sour cream
- 3 tablespoons of flour
- Salt and pepper

Method:

1. Wash the nettle leaves, then blanch them and drain them. This will get rid of the sting.
2. Put the nettles and spinach into a saucepan and pour the hot stock over them.
3. Season with salt and pepper and simmer for 45 minutes, adding more liquid if needed.
4. Let cool and blend.

5. Mix the flour with some milk to create a smooth paste, and add it to a saucepan with the blended soup. Bring to a boil.
6. Chop the sausages into small rounds and add to the soup.
7. Swirl in the sour cream and serve immediately.

Main Courses

To really show appreciation for the wonders of nature, you can make foraged foods a part of your main meals. They can become a centerpiece of the dish or complement the other flavors in a unique and delicious way.

Burdock Shitake Fried Rice

This healthy rice dish is delicious, combining the earthy flavors of burdock root with the umami taste of shiitake.

Ingredients:

- 3 cups oil pre-cooked rice
- 3 tablespoons of coconut oil
- 1 large onion, diced
- 3 cloves of garlic, minced
- ½ teaspoon of fresh ginger, minced
- 1 medium carrot, peeled and chopped into small pieces
- ¼ cup of frozen peas
- 5 shiitake mushrooms, chopped into small pieces
- 1 4-inch burdock root, chopped into small pieces
- 2 eggs
- 2 to 3 tablespoons of soy sauce
- 1 scallion, thinly sliced

Method:

1. Presoak burdock in water for 20 minutes. Drain, rinse with cold water, then dry. Fry in 1 tablespoon of oil until it's soft and set aside.
2. Beat the eggs in a small bowl and set aside.
3. Heat the remaining oil in a wok over medium heat. Add the onion and cook until translucent.
4. Lower the heat and add the garlic, ginger, and carrots. Stir until the carrots are slightly softened.
5. Add peas and stir, then add the mushrooms and the burdock root and stir for a minute.
6. Push the vegetables to one side and tilt the wok so that the oil slides to the empty side. If need be, add more oil. Pour in the beaten eggs and scramble.
7. Add the rice and soy sauce, then mix everything together. Stir for 2-3 minutes to warm the rice.
8. Garnish with the scallion and serve right away.

Notes:

- If you have wild mushrooms, use them instead or as well as the shitake.
- If using dried mushrooms, soak in water before using to rehydrate them, then use as normal.

Yarrow Omelet

Simple, tasty, and healthy, this omelet makes the perfect breakfast or lunch in minutes.

Ingredients:

- 6 eggs
- ¼ cup yarrow, finely chopped
- 1 small onion, finely chopped
- Butter or oil
- Salt and pepper

Method:

1. Beat the eggs in a mixing bowl. Add the yarrow and onion and combine.
2. Heat the butter or oil in a frying pan on low heat and add the eggs.
3. Move the pan around to allow the egg to fill the entire pan and cook the eggs, slightly disturbing them as you go.
4. When the egg has just set, serve with salt and pepper.

Dips and Dressings

Sauces and dressings are an integral and often overlooked part of a great meal. Make them shine with your foraged goodies.[118]

Japanese Knotweed Dip

This slightly spicy dip is perfect with celery and carrots. You can also use it as a salad dressing, so it's handy to keep around.

Ingredients:

- 1½ cups of Japanese knotweed, chopped
- ½ tablespoon of coconut oil
- ¾ cup of olive oil
- ½ cup of red wine vinegar
- 2-3 cloves of fresh garlic
- ½ teaspoon of cayenne pepper (or more)
- ½ teaspoon of salt

Method:

1. Sauté the Japanese knotweed in the coconut oil until soft. When softened, remove from the heat and let cool.

---

[118] (*Wild Food Dressing, Dip & Vinaigrette Recipes*, n.d.)

2. Put all of the ingredients in a food processor and blend well.
3. It can be used as a dressing right away or refrigerated. It will thicken overnight. Store for a week.

To Conclude

I've just scratched the surface of some of the great recipes you can make with your foraged ingredients. With a little bit of imagination, you can revolutionize your cooking and make it healthier, cheaper, and more delicious. Now we just have a few things to wrap up in the final chapter.

Leave a Review

I hope you have enjoyed reading this book. If you did, it would mean the world to me if you could spare 60 seconds to write a brief review of the book on Amazon, it doesn't have to be long!

Conclusion

So, you've almost made it through this guide. By now, you should have a solid foundation of foraging knowledge and tips to help you to get out there and start exploring the wild edibles near you. Here's what we've learned:

- Even if you live in a modern, technology-driven world with a lifestyle to match, foraging is still a practical and beneficial skill to develop.
- A little botany knowledge can give you a fantastic head-start as a beginner forager.
- Safe and responsible foraging is better for you and the environment alike. A forager is a guardian of the natural world.
- Following simple guidelines will improve your foraging experience.
- Forage with friends or family, and find other foragers to give you local tips.
- Only eat something that you can reliably identify.
- Foraging can provide a wide range of plants and mushrooms, all of which you can use to create delicious and nutritious meals.
- Preserving your foraged food can help you enjoy them all year round, no matter what's in season.

Millenia ago, foraging was humanity's only means of survival. But just because society has moved away from this lifestyle, it doesn't mean that foraging should fall by the wayside. With the proper knowledge and skills by your side, you can turn this skill into a sustainable and rewarding addition to your way of life. Teach yourself and your children to be more self-sufficient and use the gifts that nature gives us.

No matter where you live or what the season is, there are always wild edibles growing nearby, just waiting for you to discover them. Take this book with you, along with some trusty tools, and get out there.

I'd love to hear your thoughts on this book, so feel free to leave a review and let me know what you think.

A Free Gift to Our Readers

Thanks for buying this book! Here is a list of 5 foolish mistakes foragers make and how to avoid them that you can read right now! Visit the link below for your free copy!

http://eddieholden.com

References

Adamant, A. (2018, March 25). *Spring Foraging ~ 20+ Wild Spring Edibles*. Practical Self Reliance. Retrieved April 19, 2022, from https://practicalselfreliance.com/spring-foraging/

Adamant, A. (2018, September 28). *Foraging Beech Nuts*. Practical Self Reliance. Retrieved April 25, 2022, from https://practicalselfreliance.com/foraging-beech-nuts/

Adamant, A. (2018, November 13). *Winter Foraging in Cold Climates: 50+ Wild Foods in the Snow*. Practical Self Reliance. Retrieved April 26, 2022, from https://practicalselfreliance.com/winter-foraging/

Albert, S. (n.d.). *Seven Ways to Prepare Horseradish*. Harvest to Table. Retrieved April 22, 2022, from https://harvesttotable.com/horseradish_the_peak_season_fo/

Allday, D. (n.d.). *Identifying, Harvesting & Cooking Bamboo | EcoFarming Daily*. Eco Farming Daily. Retrieved April 14, 2022, from https://www.ecofarmingdaily.com/grow-crops/grow-fruits-vegetables/fruit-and-vegetable-crops/identifying-harvesting-cooking-bamboo/

American Beech. (n.d.). SBVPA. Retrieved April 25, 2022, from https://sbvpa.org/treetrail/american-beech/

American chestnut. (n.d.). Wikipedia. Retrieved April 25, 2022, from https://en.wikipedia.org/wiki/American_chestnut

Apple (Malus x domestica) - British Trees. (n.d.). Woodland Trust. Retrieved April 25, 2022, from https://www.woodlandtrust.org.uk/trees-woods-and-wildlife/british-trees/a-z-of-british-trees/apple/

Arnarson, A. (2019, May 8). *Apples 101: Nutrition Facts and Health Benefits.* Healthline. Retrieved April 25, 2022, from https://www.healthline.com/nutrition/foods/apples#vitamins-and-minerals

Baker, S. (2021, December 28). *What We Can Learn from Foraging, Other than How to Forage.* Impakter. Retrieved March 22, 2022, from https://impakter.com/what-we-can-learn-from-foraging-other-than-how-to-forage/

Baptisia, a Wild Asparagus Look Alike. (n.d.). Forager Chef. Retrieved April 15, 2022, from https://foragerchef.com/baptisia-a-wild-asparagus-look-alike/

Basic Botany. (n.d.). UF/IFAS Extension. Retrieved April 6, 2022, from https://sfyl.ifas.ufl.edu/media/sfylifasufledu/wakulla/images/master-gardeners/Botany-101.pdf

Basics of Plant Morphology. (n.d.). American Botanical Council. Retrieved April 7, 2022, from https://www.herbalgram.org/resources/medicinal-plant-id/background/basics-of-plant-morphology/

Bauer, E. (n.d.). *Cranberry Sauce Recipe.* Simply Recipes. Retrieved April 26, 2022, from https://www.simplyrecipes.com/recipes/cranberry_sauce/

Beechnut - dried Nutrition Facts | Calories in Beechnut - dried. (n.d.). CheckYourFood. Retrieved April 25, 2022, from https://www.checkyourfood.com/ingredients/ingredient/74/beechnut-dried

A beginner's guide to autumn foraging | Live Better. (2014, September 19). The Guardian. Retrieved April 25, 2022, from https://www.theguardian.com/lifeandstyle/2014/sep/19/beginners-guide-to-autumn-foraging

The beginner's guide to late summer foraging. (2011, August 2). The Ecologist. Retrieved April 21, 2022, from https://theecologist.org/2011/aug/02/beginners-guide-late-summer-foraging

Berries that Looks Like Blueberries » Top Suggestions. (n.d.). Garden.eco. Retrieved April 14, 2022, from https://www.garden.eco/berries-that-look-like-blueberries

blackberry | fruit | Britannica. (n.d.). Encyclopedia Britannica. Retrieved April 25, 2022, from https://www.britannica.com/plant/blackberry-fruit

Blackberry Leaf Tea: A Herbal Remedy For Your Health. (2021, June 24). Backyard Garden Lover. Retrieved April 25, 2022, from https://www.backyardgardenlover.com/blackberry-leaf-tea/

Black Walnut (Juglans nigra): Benefits, Supplements, and Safety. (2019, March 29). Healthline. Retrieved April 26, 2022, from https://www.healthline.com/nutrition/black-walnut#intro

Black Walnut (Juglans nigra) - British Trees. (n.d.). Woodland Trust. Retrieved April 26, 2022, from https://www.woodlandtrust.org.uk/trees-woods-and-wildlife/british-trees/a-z-of-british-trees/black-walnut/

Blankespoor, J., & Gemma, M. (2021, April 13). *Essential Foraging Tools and Supplies.* Chestnut School of Herbal Medicine. Retrieved March 27, 2022, from https://chestnutherbs.com/essential-foraging-tools-and-supplies/

Blueberry Tea: Is It Good for You? Pros and Cons, Nutrition Information, and More. (2020, November 16). WebMD. Retrieved April 14, 2022, from https://www.webmd.com/diet/health-benefits-blueberry-tea#1

Boletus edulis, Cep, Penny Bun Bolete mushroom. (n.d.). First Nature. Retrieved April 27, 2022, from https://www.first-nature.com/fungi/boletus-edulis.php

bract | plant structure | Britannica. (n.d.). Encyclopedia Britannica. Retrieved April 18, 2022, from https://www.britannica.com/science/bract-plant-structure

Brennan, D. (2020, August 31). *Blackberries: Health Benefits, Nutrients, Preparation, and More.* WebMD. Retrieved April 25, 2022, from https://www.webmd.com/diet/health-benefits-blackberries

Brennan, D. (2020, September 2). *Chestnuts: Health Benefits, Nutrients per Serving, Preparation Information, and More.* WebMD. Retrieved April 25, 2022, from https://www.webmd.com/diet/health-benefits-chestnuts

Brennan, D. (2021, March 30). *Does Vitamin D Boost Mental Health?* WebMD. Retrieved March 26, 2022, from https://www.webmd.com/vitamins-and-supplements/what-to-know-about-vitamin-d-and-mental-health

Brenner, L. (2018, April 17). *How to Identify Wild Mushrooms in Florida.* Sciencing. Retrieved April 28, 2022, from https://sciencing.com/identify-wild-mushrooms-florida-6329232.html

Breyer, M. (n.d.). *7 Simple Recipes for Preserving Edible Flowers.* Treehugger. Retrieved May 1, 2022, from https://www.treehugger.com/basic-recipes-preserve-edible-flowers-4857763

Bryant, C. W. (n.d.). *What is the universal edibility test? | HowStuffWorks.* Adventure | HowStuffWorks. Retrieved April 12, 2022, from https://adventure.howstuffworks.com/universal-edibility-test.htm

Burdock Root and Miso Soup Recipe. (n.d.). Edible Wild Food. Retrieved May 2, 2022, from https://www.ediblewildfood.com/burdock-root-and-miso-soup.aspx

Chickweed: Benefits, Side Effects, Precautions, and Dosage. (2020, April 21). Healthline. Retrieved April 20, 2022, from https://www.healthline.com/nutrition/chickweed-benefits#downsides

Chickweed Pesto: Wild Greens Superfood Recipe. (2018, March 23). Grow Forage Cook Ferment. Retrieved April 21, 2022, from https://www.growforagecookferment.com/chickweed-pesto/

Chicory. (n.d.). Wikipedia. Retrieved April 22, 2022, from https://en.wikipedia.org/wiki/Chicory#Nutrition

The Classification Of Plants - Annuals, Biennials and Perennials. (n.d.). Byjus. Retrieved April 7, 2022, from https://byjus.com/biology/the-classification-of-plants/

Coelho, S. (n.d.). *50 Edible Wild Plants You Can Forage for a Free Meal.* MorningChores. Retrieved April 13, 2022, from https://morningchores.com/edible-wild-plants/

Dandelion | Garden Organic. (2022, April 4). Garden Organic |. Retrieved April 18, 2022, from https://www.gardenorganic.org.uk/weeds/dandelion

Davis, E. M. (2015, August 24). *A tale of four daisies.* Wild Food Girl. Retrieved April 18, 2022, from https://wildfoodgirl.com/2015/a-tale-of-four-daisies/

Deane. (2011, 08 30). *Why Forage?* Eat the Weeds and Other Things, Too. https://www.eattheweeds.com/why-forage/

Deane, G. (n.d.). *Apples, Wild Crabapples - Eat The Weeds and other things, too*. Eat the Weeds. Retrieved April 25, 2022, from https://www.eattheweeds.com/apples-wild-crabapples/

Demers, A. (2021, February 6). *The Story of the American Chestnut*. Eat The Planet. Retrieved April 25, 2022, from https://eattheplanet.org/the-story-of-the-american-chestnut/

Demers, A. (2021, December 13). *Dulse: A Seafaring Superfood*. Eat The Planet. Retrieved April 22, 2022, from https://eattheplanet.org/dulse-a-seafaring-superfood/

Docio, A. (n.d.). *Garlic mustard: Foraging for culinary and medicinal use*. British Local Food. Retrieved April 18, 2022, from https://britishlocalfood.com/garlic-mustard/

Docio, A. (n.d.). *What is foraging and what do foragers eat? - BritishLocalFood*. British Local Food. Retrieved March 11, 2022, from https://britishlocalfood.com/what-is-foraging/

Eating Invasive Bastard Cabbage for the First Time. (2016, March 23). The Foraged Foodie. Retrieved April 15, 2022, from https://foragedfoodie.blogspot.com/2016/03/eating-invasive-bastard-cabbage.html

Edible Plants Guide. (n.d.). Survival Manual. Retrieved April 13, 2022, from https://www.survival-manual.com/edible-plants/edible-plant-guide.php

11 Edible Mushrooms in the US (And How to Tell They're Not Toxic). (2018, December 4). PlantSnap. Retrieved April 27, 2022, from https://www.plantsnap.com/blog/edible-mushrooms-united-states/

Field Garlic Information and Facts. (n.d.). Specialty Produce. Retrieved April 22, 2022, from https://specialtyproduce.com/produce/Field_Garlic_17295.php

Fischer, D. W. (n.d.). *Poisonous American Mushrooms - AmericanMushrooms.com*. American Mushrooms. Retrieved April 27, 2022, from https://www.americanmushrooms.com/toxicms.htm

Fishman, E., & Jones, P. (2020, September 14). *54 Mushroom Recipes So Good, They're Magic*. Bon Appetit. Retrieved April 28, 2022, from https://www.bonappetit.com/recipes/slideshow/mushroom-recipes-slideshow

5 Emerging Benefits and Uses of Yarrow Tea. (2019, December 12). Healthline. Retrieved April 22, 2022, from https://www.healthline.com/nutrition/yarrow-tea#How-to-add-it-to-your-diet

Foraging 101: How to identify poisonous plants in the wild (and in your garden). (n.d.). Survival news. https://survival.news/2020-06-17-foraging-identify-poisonous-plants-wild-and-garden.html

Foraging Calendar. (n.d.). Meadows and More. Retrieved March 27, 2022, from https://www.meadowsandmore.com/in-the-field/foraging-calendar/

Foraging for Chicory. (2016, October 6). Grow Forage Cook Ferment. Retrieved April 22, 2022, from https://www.growforagecookferment.com/foraging-for-chicory/

Foraging Guide Raspberry | UK Foraging. (n.d.). The Foraging Course Company. Retrieved April 21, 2022, from https://www.foragingcoursecompany.co.uk/foraging-guide-raspberry

Foraging in Early Spring: Wild Edible Plants to Gather Now — Good Life Revival. (2021, April 20). Good Life Revival. Retrieved April 20, 2022, from https://thegoodliferevival.com/blog/foraging-spring-wild-edible-plants

Foraging Juniper Berries for Food and Medicine. (2020, November 23). Grow Forage Cook Ferment. Retrieved April 26, 2022, from https://www.growforagecookferment.com/foraging-for-juniper-berries/

Foraging Legality — Four Season Foraging. (2019, April 1). Four Season Foraging. Retrieved March 27, 2022, from https://www.fourseasonforaging.com/blog/2019/3/19/foraging-legality

Foraging Plantain: Identification and Uses. (2020, July 28). Grow Forage Cook Ferment. Retrieved April 20, 2022, from https://www.growforagecookferment.com/plantain-natures-band-aid/

Foraging: Ultimate Guide to Wild Food. (n.d.). Wild Edible. Retrieved March 27, 2022, from https://www.wildedible.com/foraging

Foraging Yarrow: Identification, Lookalikes, and Uses. (2021, February 15). Grow Forage Cook Ferment. Retrieved April 22, 2022, from https://www.growforagecookferment.com/foraging-for-yarrow/

40 Most Common Edible Wild Plants in North America. (2015, May 22). Preparedness Advice. Retrieved April 13, 2022, from https://preparednessadvice.com/40-common-edible-wild-plants-north-america/

4 Essential Foraging Tools. (2019, July 22). Rustic Farm Life. Retrieved March 27, 2022, from https://www.rusticfarmlife.com/essential-foraging-tools/

Foxglove (Digitalis purpurea) - British Plants. (n.d.). Woodland Trust. Retrieved April 11, 2022, from https://www.woodlandtrust.org.uk/trees-woods-and-wildlife/plants/wild-flowers/foxglove/

Fratt, K. (2018, August 6). *The Forager's Guide to Plant Identification*. PlantSnap. Retrieved April 7, 2022, from https://www.plantsnap.com/blog/the-foragers-guide-to-plant-identification/

Genus Arundinaria: Native bamboo of North America. (2020, June 18). Bambu Batu. Retrieved April 14, 2022, from https://bambubatu.com/native-bamboo-of-north-america/

Grant, B. L. (2020, November 16). *Prickly Pear Fruit Harvest - Information On Picking Prickly Pear Fruit*. Gardening Know How. Retrieved April 14, 2022, from https://www.gardeningknowhow.com/ornamental/cacti-succulents/prickly-pear/harvesting-prickly-pear-fruit.htm

Grant, B. L. (2021, February 27). *Is It Safe To Pick Juniper Berries – Learn About Harvesting Juniper Berries*. Gardening Know How. Retrieved April 26, 2022, from https://www.gardeningknowhow.com/ornamental/shrubs/juniper/juniper-berry-harvesting.htm

Greaves, V. (2021, January 13). *How to Store Berries to Keep Them Fresh*. Allrecipes. Retrieved April 30, 2022, from https://www.allrecipes.com/article/how-to-store-berries/

Ground-elder - Aegopodium podagraria. (n.d.). Roots to Health. Retrieved April 20, 2022, from https://www.rootstohealth.co.uk/herbs/ground-elder/

Ground elder - characteristics, cultivation, care and use. (2019, July 18). live-native.com. Retrieved April 20, 2022, from https://www.live-native.com/ground-elder/

Ground elder - characteristics, cultivation, care and use. (2019, July 18). live-native.com. Retrieved April 20, 2022, from https://www.live-native.com/ground-elder/

Guide, S. (n.d.). *Basic Jam Recipe*. Martha Stewart. Retrieved April 30, 2022, from https://www.marthastewart.com/1128024/basic-jam-recipe

Harvesting and Processing Edible Wild Plants. (n.d.). Ontario Nature. Retrieved April 28, 2022, from https://ontarionature.org/wp-content/uploads/2017/10/Ontario_Nature_Harvesting_and_Processing_Edible_Wild_Plants_Best_Practices_Guide.pdf

Haughton, C. S. (n.d.). *Wild Carrot.* Weed Science Society of America. Retrieved April 22, 2022, from https://wssa.net/wp-content/themes/WSSA/WorldOfWeeds/wildcarrot.html

Hodgkins, K. (2021, April 10). *Poisonous Plants Common in USA [Full List With Pictures].* Greenbelly Meals. Retrieved April 10, 2022, from https://www.greenbelly.co/pages/poisonous-plants-identification-guide

Horseradish - A Foraging Guide to Its Food, Medicine and Other Uses. (n.d.). Eatweeds. Retrieved April 22, 2022, from https://www.eatweeds.co.uk/horseradish-armoracia-rusticana

House, K. M. (n.d.). *7 Plants To Forage In Summer.* Self-Reliance. Retrieved April 22, 2022, from https://www.self-reliance.com/2016/06/7-plants-to-forage-in-summer/

How nature benefits mental health. (n.d.). Mind. Retrieved March 27, 2022, from https://www.mind.org.uk/information-support/tips-for-everyday-living/nature-and-mental-health/how-nature-benefits-mental-health/

How to cook mushrooms - you won't believe the secret! (2020, March 2). Imagelicious.com. Retrieved April 28, 2022, from https://www.imagelicious.com/blog/how-to-cook-perfect-mushrooms

How to Harvest and Use Rose Hips - Flowers. (2021, October 7). The Spruce. Retrieved April 25, 2022, from https://www.thespruce.com/what-are-rose-hips-and-what-do-they-do-1403046

How to Identify Field Garlic - Foraging for Wild Edible Plants — Good Life Revival. (2017, March 21). Good Life Revival. Retrieved April 22, 2022, from https://thegoodliferevival.com/blog/wild-field-garlic

How to Make Easy Homemade Jelly: Basic Jelly Recipe With Tips, Ingredients, and Flavors - 2022. (2021, August 2). MasterClass. Retrieved April 30, 2022, from https://www.masterclass.com/articles/how-to-make-easy-homemade-jelly-basic-jelly#classic-concord-grape-jelly-recipe

How To Preserve Fresh Herbs: Tips & Tricks for Storing. (2021, June 4). The Markets At Shrewsbury. Retrieved April 29, 2022, from https://www.marketsatshrewsbury.com/blog/preserve-fresh-herbs/

How to Use Lambsquarter from Root to Plant to Seed. (n.d.). Chelsea Green Publishing. Retrieved April 18, 2022, from https://www.chelseagreen.com/2021/use-lambsquarter-from-root-to-seed/

Huffstetler, E. (2019, November 20). *Foraging for Food: a Monthly Guide*. LiveAbout. Retrieved March 31, 2022, from https://www.liveabout.com/foraging-for-food-a-monthly-guide-1388185

Huffstetler, E. (2021, July 23). *How to Harvest and Store Chestnuts*. The Spruce Eats. Retrieved April 25, 2022, from https://www.thespruceeats.com/how-to-harvest-and-store-chestnuts-1388176

Junipers - Eat The Weeds and other things, too. (n.d.). Eat the Weeds. Retrieved April 26, 2022, from https://www.eattheweeds.com/junipers/

Kubala, J. (2021, June 3). *Foraging for Food: Tips, Common Foods, Safety, and More*. Healthline. Retrieved April 13, 2022, from https://www.healthline.com/nutrition/foraging-for-food#tips-for-beginners

Larum, D. (2021, September 15). *Is Yarrow Good For You: Medicinal, Edible, And Herbal Yarrow Plants*. Gardening Know How. Retrieved April 22, 2022, from https://www.gardeningknowhow.com/edible/herbs/yarrow/yarrow-plant-uses-and-benefits.htm

Lee, C. (2022, April 19). *Storing and Preserving Wild Mushrooms | 3 Effective Methods*. VENCHAS. Retrieved April 28, 2022, from https://venchas.com/storing-wild-mushrooms/

Lee, L. W. (2017, July 18). *Does Light Affect How Quickly Foods Spoil?* Healthfully. Retrieved April 28, 2022, from https://healthfully.com/523460-does-light-affect-how-fast-foods-spoil.html

Licavoli, K. (2021, May 26). *Universal Edibility Test: THE Complete Guide [Step-by-Step]*. Greenbelly Meals. Retrieved April 12, 2022, from https://www.greenbelly.co/pages/universal-edibility-test

Linnaeus, C. (n.d.). *Allium vineale*. Wikipedia. Retrieved April 22, 2022, from https://en.wikipedia.org/wiki/Allium_vineale

Linnaeus, C. (n.d.). *Cicuta*. Wikipedia. Retrieved April 10, 2022, from https://en.wikipedia.org/wiki/Cicuta

Linnaeus, C. (n.d.). *Daucus carota*. Wikipedia. Retrieved April 22, 2022, from https://en.wikipedia.org/wiki/Daucus_carota#Beneficial_weed

Linnaeus, C. (n.d.). *Diospyros virginiana*. Wikipedia. Retrieved April 25, 2022, from https://en.wikipedia.org/wiki/Diospyros_virginiana

Linnaeus, C. (n.d.). *Horseradish*. Wikipedia. Retrieved April 22, 2022, from https://en.wikipedia.org/wiki/Horseradish

Linnaeus, C. (n.d.). *Juniperus communis*. Wikipedia. Retrieved April 26, 2022, from https://en.wikipedia.org/wiki/Juniperus_communis

Linnaeus, C. (n.d.). *Rubus occidentalis*. Wikipedia. Retrieved April 21, 2022, from https://en.wikipedia.org/wiki/Rubus_occidentalis

Linnaeus, C. (n.d.). *Urtica dioica*. Wikipedia. Retrieved April 6, 2022, from https://en.wikipedia.org/wiki/Urtica_dioica

Logiste, D. (2015, July 23). *The Foragers Bible - How To Store Edible Wild Plants*. Homemade Recipes. Retrieved April 30, 2022, from https://homemaderecipes.com/the-foragers-bible-how-to-store-edible-wild-plants/

Long, L. (2019, September 15). *Yarrow Benefits and How to make yarrow Tea?* Orchards Near Me. Retrieved April 22, 2022, from https://orchardsnearme.com/2019/09/15/yarrow-benefits-and-how-to-make-yarrow-tea/

Mabey, R. (2015, August 14). *13 reasons you need to start foraging*. Healthista. Retrieved March 22, 2022, from https://www.healthista.com/13-reasons-to-be-outdoors-and-foraging-for-food/

MacWelch, T. (2019, August 13). *13 Toxic Wild Plants That Look Like Food*. Outdoor Life. Retrieved April 26, 2022, from https://www.outdoorlife.com/13-toxic-wild-plants-not-food/

Melanthiaceae – Death Camas – Better Learning Through Botany. (n.d.). Better Learning Through Botany. Retrieved April 11, 2022, from https://willamettebotany.org/melanthiaceae-death-camas/

Meredith, L. (2014, December 4). *Cold-Weather Foraging for Wild, American Persimmons.* Mother Earth News. Retrieved April 25, 2022, from https://www.motherearthnews.com/real-food/foraging-for-wild-american-persimmons-zbcz1412/

Miller, L. (2020, September 28). *Canning Fresh Mushrooms: How To Can Mushrooms From The Garden.* Gardening Know How. Retrieved April 29, 2022, from https://www.gardeningknowhow.com/edible/vegetables/mushrooms/home-canning-mushrooms.htm

Miller, P. (n.d.). *Malus.* Wikipedia. Retrieved April 25, 2022, from https://en.wikipedia.org/wiki/Malus

Monthly foraging guide: what's in season, where to find it, and how to forage responsibly. (n.d.). Countryfile.com. Retrieved March 27, 2022, from https://www.countryfile.com/how-to/foraging/monthly-foraging-guide-whats-in-season-where-to-find-it-and-how-to-forage-responsibly/

Mullins, A. (2014, September 12). *Top 10 Things to Forage in Autumn.* And Here We Are. Retrieved April 24, 2022, from https://andhereweare.net/top-10-things-to-forage-in-autumn/

Myers, V. R. (2022, March 25). *Five Common Varieties of Beech Trees.* The Spruce. Retrieved April 25, 2022, from https://www.thespruce.com/five-kinds-of-beech-trees-3269706

Nagy, T. (2015, February 9). *Foraging Fun: Highbush Cranberries - The Permaculture Research Institute.* Permaculture Research Institute. Retrieved April 26, 2022, from https://www.permaculturenews.org/2015/02/09/foraging-fun-highbush-cranberries/

Narrow leaf plantain facts and health benefits. (n.d.). Health Benefits Times. Retrieved April 20, 2022, from https://www.healthbenefitstimes.com/narrow-leaf-plantain/

Neuharth, S., & Sammak, H. (2019, April 15). *The Total Guide to Morel Mushroom Hunting.* MeatEater. Retrieved April 27, 2022, from https://www.themeateater.com/cook/foraging/the-meateater-guide-to-morel-mushroom-hunting

Neuharth, S., & Sammak, H. (2020, August 27). *The 12 Best Edible Wild Mushrooms*. MeatEater. Retrieved April 27, 2022, from https://www.themeateater.com/cook/foraging/the-12-best-edible-fall-mushrooms

Neverman, L. (n.d.). *Broadleaf Plantain – The "Weed" You Won't Want to Be Without*. Common Sense Home. Retrieved April 20, 2022, from https://commonsensehome.com/broadleaf-plantain/#Nutrient_Dense_Food

Orr, E. (n.d.). *Foraging Reference: 130+ Edible Wild Plants*. Wild Edible. Retrieved April 13, 2022, from https://www.wildedible.com/foraging-reference

Overhiser, S. (2020, August 17). *How to Clean Mushrooms...the Right Way – A Couple Cooks*. A Couple Cooks. Retrieved April 28, 2022, from https://www.acouplecooks.com/how-to-clean-mushrooms-the-right-way/

Palomo, E. (n.d.). *How to Get Seeds From Queen Anne's Lace Plants*. Garden Guides. Retrieved April 22, 2022, from https://www.gardenguides.com/92683-seeds-queen-annes-lace-plants.html

Parker, J. (2020, April 3). *5 Ways To Preserve Fruits and Vegetables in Any Season*. Mother Of Health. Retrieved April 30, 2022, from https://motherofhealth.com/preserve-fruits-and-vegetables

Plant Morphology. (n.d.). Biocyclopedia. Retrieved April 7, 2022, from https://biocyclopedia.com/index/plant_morphology.php

Poisonous Plants: Geographic Distribution | NIOSH. (n.d.). CDC. Retrieved April 11, 2022, from https://www.cdc.gov/niosh/topics/plants/geographic.html

Prendergast, A. (n.d.). *Dandelions – Botanical Society of Britain & Ireland*. BSBI. Retrieved April 6, 2022, from https://bsbi.org/identification/taraxacum

Queen Anne's Lace - The Wild Carrot. (n.d.). The World Carrot Museum. Retrieved April 22, 2022, from http://www.carrotmuseum.co.uk/wild.html

Raab, C. (n.d.). *Food Safety & Preservation: Herbs and Vegetables in Oil*. Cornell Cooperative Extension. Retrieved April 28, 2022, from http://ccetompkins.org/resources/herbs-vegetables-in-oil

Radiant Rose Hips: How to Harvest, Dry and Use Rosehips. (2022, February 9). Homestead and Chill. Retrieved April 25, 2022, from https://homesteadandchill.com/rose-hips-harvest-dry-use/

Red Raspberries: Nutrition Facts, Benefits and More. (2018, October 13). Healthline. Retrieved April 21, 2022, from https://www.healthline.com/nutrition/raspberry-nutrition

Retta, M. (2021, December 27). *How TikTok Foragers Are Making Sustainability Accessible.* Teen Vogue. Retrieved March 7, 2022, from https://www.teenvogue.com/story/how-tiktok-foragers-are-making-sustainability-accessible

Rey, E. (n.d.). *Common Yarrow (Achillea millefolium).* USDA Forest Service. Retrieved April 22, 2022, from https://www.fs.fed.us/wildflowers/plant-of-the-week/achillea_millefolium.shtml

Ribwort Plantain, Narrow leaf Plantain, Plantago lanceolata. (n.d.). Wild Food UK. Retrieved April 20, 2022, from https://www.wildfooduk.com/edible-wild-plants/ribwort-plantain/

Rosa acicularis. (n.d.). Wikipedia. Retrieved April 25, 2022, from https://en.wikipedia.org/wiki/Rosa_acicularis

Rose Hips: When, How, and Why to Harvest. (2016, August 25). Backyard Forager. Retrieved April 25, 2022, from https://backyardforager.com/rose-hips-harvest-process/

Rose Hip Syrup: Foraged and Made with Honey. (2020, October 24). Grow Forage Cook Ferment. Retrieved April 25, 2022, from https://www.growforagecookferment.com/rose-hip-syrup/

Rubus strigosus. (n.d.). Wikipedia. Retrieved April 21, 2022, from https://en.wikipedia.org/wiki/Rubus_strigosus

Russell, P. (n.d.). *Canning Fruit : 10 Steps (with Pictures).* Instructables. Retrieved April 30, 2022, from https://www.instructables.com/Canning-Fruit/

Sarkar, J. (2012, December 29). *Aegopodium podagraria (Ground elder).* Only Foods. Retrieved April 20, 2022, from https://www.onlyfoods.net/aegopodium-podagraria-ground-elder.html

Sayer, A. (2018, June 7). *How to Safely Forage for Mushrooms • New Life On A Homestead.* New Life On A Homestead. Retrieved April 28, 2022, from https://www.newlifeonahomestead.com/how-to-forage-for-mushrooms/

Schatz, T. (2021, April 21). *Wild Edibles of Spring: The Fine Art of Foraging for Food*. Back Road Ramblers. Retrieved April 20, 2022, from https://backroadramblers.com/spring-wild-edibles-for-camping/

Schipani, S. (2019, May 30). *How to forage responsibly*. Hello Homestead. Retrieved April 13, 2022, from https://hellohomestead.com/how-to-forage-responsibly/

Shaw, H. (2010, December 14). *Black Walnuts - Harvesting and Cooking Black Walnuts | Hank Shaw*. Hunter Angler Gardener Cook. Retrieved April 26, 2022, from https://honest-food.net/black-walnuts-and-holiday-cheer/

6 Surprising Ways Summer Can Be Good For Your Health. (2020, July 24). Lull. Retrieved April 21, 2022, from https://lull.com/blog/6-ways-summer-can-be-good-for-your-health/

Sjdin, M. S. (n.d.). *Gourmet Mushroom Risotto Recipe*. Allrecipes. Retrieved April 28, 2022, from https://www.allrecipes.com/recipe/85389/gourmet-mushroom-risotto/

Soken, E. (2020, November 11). *The Best Food Preservation Methods*. Gildshire Magazines. Retrieved April 28, 2022, from https://www.gildshire.com/the-best-food-preservation-methods/

Steaven, D. (2018, January 4). *Foraging Fatality Statistics 2016 (Please Share)*. Eat The Planet. Retrieved April 7, 2022, from https://eattheplanet.org/foraging-fatality-statistics-2016/

Subspecies, Varieties and Cultivars – Succulent Gardens. (2017, May 22). Succulent Gardens. Retrieved April 6, 2022, from https://sgplants.com/blogs/news/subspecies-varieties-and-cultivars

Sugar as a preservative. (2021, October 20). Sugar Nutrition Resource Centre. Retrieved April 28, 2022, from https://www.sugarnutritionresource.org/news-articles/sugar-as-a-preservative

Suwak, M. (2018, March 2). *How to Harvest Wild Berries: Foraging for Beginners*. Gardener's Path. Retrieved April 21, 2022, from https://gardenerspath.com/how-to/beginners/beginner-foraging-berries/#raspberry

Thayer, S. (n.d.). *Highbush Cranberry: A Loser in the Name Game. An Edible Plants Article by Sam Thayer*. Wildflowers and Weeds. Retrieved April 26, 2022, from http://www.wildflowers-and-weeds.com/Edible_Plants/Articles/Highbush_Cranberry.htm

This is how you forage mushrooms safely (anyone can do it) – Mushroom grove. (n.d.). Mushroom grove. Retrieved April 28, 2022, from https://mushroomgrove.com/foraging-mushrooms-safely/

Top 7 Health and Nutrition Benefits of Persimmon. (2018, May 5). Healthline. Retrieved April 25, 2022, from https://www.healthline.com/nutrition/persimmon-nutrition-benefits#TOC_TITLE_HDR_2

Toxicodendron diversilobum. (n.d.). Wikipedia. Retrieved April 11, 2022, from https://en.wikipedia.org/wiki/Toxicodendron_diversilobum

TyrantFarms. (2019, June 12). *Beginner's guide to foraging: 12 rules to follow.* TyrantFarms. https://www.tyrantfarms.com/beginners-guide-to-foraging-rules-to-follow/

Ultimate Guide to Wild Edibles: Spring Wild Edibles. (2020, May 25). Outdoor Adventure Sampler. Retrieved April 21, 2022, from https://outdooradventuresampler.com/ultimate-guide-to-wild-edibles-spring-wild-edibles/

USDA Plants Database. (n.d.). *Aegopodium podagraria L.* Plants.sc.egov.usda.gov. https://plants.sc.egov.usda.gov/home/plantProfile?symbol=AEPO

VanDerZanden, A. M. (n.d.). *Botany Basics.* Oregon State University's Professional and Continuing Education. Retrieved April 3, 2022, from https://pace.oregonstate.edu/courses/sites/default/files/resources/pdf/ch01_botany.pdf

Vinskofski, S. (2018, April 20). *The Forager's Guide to Plant Identification.* Learning and Yearning. Retrieved April 7, 2022, from https://learningandyearning.com/the-foragers-guide-to-plant-identification/

Waddington, E. (2020, March 25). *25 Edible Wild Plants To Forage For In Early Spring.* Rural Sprout. Retrieved April 19, 2022, from https://www.ruralsprout.com/edible-wild-plants-spring/

Weekes, B. (n.d.). *Picking and preserving the wild plum.* Backwoods Home Magazine. Retrieved April 21, 2022, from https://www.backwoodshome.com/picking-and-preserving-the-wild-plum/

What Is Foraging? (2021, March 29). Survive the Wild. Retrieved March 11, 2022, from https://www.survivethewild.net/learn/foraging/

What to Forage in Fall: 30+ Edible and Medicinal Plants and Mushrooms. (2018, October 5). Grow Forage Cook Ferment. Retrieved April 24, 2022, from https://www.growforagecookferment.com/what-to-forage-in-fall/

What to Forage in Winter: 30+ Edible and Medicinal Plants and Fungi. (2020, December 13). Grow Forage Cook Ferment. Retrieved April 26, 2022, from https://www.growforagecookferment.com/what-to-forage-in-winter/

White Pine Vinegar — Four Season Foraging. (2018, February 5). Four Season Foraging. Retrieved April 26, 2022, from https://www.fourseasonforaging.com/blog/2018/1/29/white-pine-vinegar

Wild Edible Plants of the Pacific Northwest. (n.d.). Northern Bushcraft. Retrieved April 14, 2022, from https://www.northernbushcraft.com/plants/

Wild Food Dessert & Snack Recipes. (n.d.). Edible Wild Food. Retrieved May 1, 2022, from https://www.ediblewildfood.com/dessert-snack-recipes.aspx

Wild Food Dressing, Dip & Vinaigrette Recipes. (n.d.). Edible Wild Food. Retrieved May 2, 2022, from https://www.ediblewildfood.com/dressing-and-dip-recipes.aspx

Wild Food: How to Forage and Store Wild Greens. (n.d.). Chelsea Green Publishing. Retrieved April 30, 2022, from https://www.chelseagreen.com/2021/how-to-forage-and-store-wild-greens/

Wild Food Salad & Sautée Recipes. (n.d.). Edible Wild Food. Retrieved May 2, 2022, from https://www.ediblewildfood.com/salad-recipes.aspx

Wild Plum facts and health benefits. (n.d.). Health Benefits Times. Retrieved April 21, 2022, from https://www.healthbenefitstimes.com/wild-plum/

Made in United States
North Haven, CT
08 March 2024